Tourism in Changing Natural Environments

Natural environments, and the human interactions that occur within, are continuously changing and evolving. This comprehensive volume explores how the impacts of climate change, natural and man-made disasters, economic instability, and other macro-environmental factors can have profound implications for local and global economies, fragile ecosystems, and human cultures and livelihoods. The authors examine the numerous ways in which changes in the natural environment impact tourism, and how the tourism industry is responding and adapting to such changes, in both developed and developing regions.

Through the various case studies that examine human interaction within what are often fragile ecosystems, this book makes it clear that, while adaptation can be passive in nature, it can and should be much more proactive, with individuals and organizations seeking improved knowledge and learning. Such actions will contribute to greater resilience within the tourism industry, whether in response to climate change and its subsequent impacts, or an increasing scarcity of the natural resources upon which tourism relies.

This book was originally published as a special issue of *Tourism Geographies*.

Natalie Ooi is Assistant Professor and Program Director of the Ski Area Management Program at Colorado State University, USA. Her areas of research include sustainable tourism, mountain resort development and impacts, social capital, and backpacker tourism.

Esther A. Duke is Western Program Director at the conservation aviation non-profit organization LightHawk, an Affiliate Faculty Member of the Human Dimensions of Natural Resource Department at Colorado State University, and a Board Member at Sonoran Joint Venture, USA. She is a social scientist and specializes in ecosystem services and building capacity for collaborative and performance-based conservation.

Joseph O'Leary is Professor in International Tourism at Colorado State University, USA. His research interests include international and domestic travel and recreation behaviour.

Tourism in Changing Natural Environments

Edited by
Natalie Ooi, Esther A. Duke and
Joseph O'Leary

LONDON AND NEW YORK

First published 2019
by Routledge
2 Park Square, Milton Park, Abingdon, Oxon, OX14 4RN

and by Routledge
52 Vanderbilt Avenue, New York, NY 10017

First issued in paperback 2020

Routledge is an imprint of the Taylor & Francis Group, an informa business

British Library Cataloguing in Publication Data
A catalogue record for this book is available from the British Library

ISBN 13: 978-0-367-67128-0 (pbk)
ISBN 13: 978-0-367-19473-4 (hbk)

Typeset in Myriad Pro
by RefineCatch Limited, Bungay, Suffolk

Publisher's Note
The publisher accepts responsibility for any inconsistencies that may have
arisen during the conversion of this book from journal articles to book chapters,
namely the possible inclusion of journal terminology.

Disclaimer
Every effort has been made to contact copyright holders for their permission to
reprint material in this book. The publishers would be grateful to hear from any
copyright holder who is not here acknowledged and will undertake to rectify
any errors or omissions in future editions of this book.

Contents

Citation Information   vii
Notes on Contributors   ix

1. Tourism in changing natural environments   1
   Natalie Ooi, Esther Duke and Joseph O'Leary

2. Costs and benefits of environmental change: tourism industry's responses
   in Arctic Finland   10
   Kaarina Tervo-Kankare, Eva Kaján and Jarkko Saarinen

3. Measuring park visitation vulnerability to climate extremes in U.S. Rockies
   National Parks tourism   32
   Theresa M. Jedd, Michael J. Hayes, Carlos M. Carrillo, Tonya Haigh,
   Christopher J. Chizinski and John Swigart

4. Climate and visitation to Utah's 'Mighty 5' national parks   58
   Jordan W. Smith, Emily Wilkins, Riana Gayle and Chase C. Lamborn

5. Weather sensitivity and climate change perceptions of tourists: a
   segmentation analysis   81
   Emily Wilkins, Sandra de Urioste-Stone, Aaron Weiskittel and Todd Gabe

6. Micro-level assessment of regional and local disaster impacts in tourist
   destinations   98
   Jürgen Schmude, Sahar Zavareh, Katrin Magdalena Schwaiger and Marion Karl

7. Exploring stakeholder groups through a testimony analysis on the
   Hawaiian aquarium trade   117
   Brooke A. Porter

Index   139

Citation Information

The chapters in this book were originally published in *Tourism Geographies*, volume 20, issue 2 (March 2018). When citing this material, please use the original page numbering for each article, as follows:

Chapter 1
Tourism in changing natural environments
Natalie Ooi, Esther Duke and Joseph O'Leary
Tourism Geographies, volume 20, issue 2 (March 2018), pp. 193–201

Chapter 2
Costs and benefits of environmental change: tourism industry's responses in Arctic Finland
Kaarina Tervo-Kankare, Eva Kaján and Jarkko Saarinen
Tourism Geographies, volume 20, issue 2 (March 2018), pp. 202–223

Chapter 3
Measuring park visitation vulnerability to climate extremes in U.S. Rockies National Parks tourism
Theresa M. Jedd, Michael J. Hayes, Carlos M. Carrillo, Tonya Haigh, Christopher J. Chizinski and John Swigart
Tourism Geographies, volume 20, issue 2 (March 2018), pp. 224–249

Chapter 4
Climate and visitation to Utah's 'Mighty 5' national parks
Jordan W. Smith, Emily Wilkins, Riana Gayle and Chase C. Lamborn
Tourism Geographies, volume 20, issue 2 (March 2018), pp. 250–272

Chapter 5
Weather sensitivity and climate change perceptions of tourists: a segmentation analysis
Emily Wilkins, Sandra de Urioste-Stone, Aaron Weiskittel and Todd Gabe
Tourism Geographies, volume 20, issue 2 (March 2018), pp. 273–289

Chapter 6
Micro-level assessment of regional and local disaster impacts in tourist destinations
Jürgen Schmude, Sahar Zavareh, Katrin Magdalena Schwaiger and Marion Karl
Tourism Geographies, volume 20, issue 2 (March 2018), pp. 290–308

Chapter 7
Exploring stakeholder groups through a testimony analysis on the Hawaiian aquarium trade
Brooke A. Porter
Tourism Geographies, volume 20, issue 2 (March 2018), pp. 309–330

For any permission-related enquiries please visit:
http://www.tandfonline.com/page/help/permissions

Notes on Contributors

Carlos M. Carrillo is a Postdoctoral Research Associate in the Department of Earth and Atmospheric Sciences at Cornell University, USA. His background is in climate variability and meteorological modeling. He received his PhD in Atmospheric Sciences from the University of Arizona.

Christopher J. Chizinski is Assistant Professor of Human Dimensions of Wildlife in the School of Natural Resources at the University of Nebraska, Lincoln, USA. He received his MS and PhD degrees from Texas Tech University. His research focuses on the socio-ecological aspects of recreational hunting and fishing.

Sandra de Urioste-Stone, PhD, is an Assistant Professor of Nature-based Tourism at the University of Maine, USA. She has carried out work on sustainable tourism, community-based tourism, human dimensions of biodiversity conservation, and sustainable development.

Esther A. Duke is Western Program Director at the conservation aviation non-profit organization LightHawk, an Affiliate Faculty Member of the Human Dimensions of Natural Resource Department at Colorado State University, and a Board Member at Sonoran Joint Venture, USA. She is a social scientist and specializes in ecosystem services and building capacity for collaborative and performance-based conservation.

Todd Gabe, PhD, is Professor of Economics at the University of Maine, USA. He teaches and conducts research in the area of regional economics, and has studied tourism-related topics such as the impacts of cruise ship passengers in Maine.

Riana Gayle is a MS student in the Department of Environment and Society at Utah State University, USA. Her research focuses on water policy in response to climate change and drought.

Tonya Haigh is a rural sociologist research specialist with the National Drought Mitigation Center in the School of Natural Resources at the University of Nebraska, Lincoln, USA. She specializes in survey research methods and agricultural vulnerability and adaptation to drought.

Michael J. Hayes is Professor in the applied climate and spatial sciences mission area within the School of Natural Resources at the University of Nebraska, Lincoln, USA. He received his MS and PhD degrees from the University of Missouri-Columbia. Dr. Hayes was with the National Drought Mitigation Center for 21 years and served as the NDMC's Director from 2007 to 2016.

Theresa M. Jedd is an environmental policy specialist and a Postdoctoral Researcher at the National Drought Mitigation Center in the School of Natural Resources at the University of Nebraska, Lincoln, USA. She received a PhD in Political Science from Colorado State University, Fort Collins, and her research seeks to understand vulnerability and adaptation in outdoor recreation under a variety of climate and weather conditions.

Eva Kaján is a Postdoctoral Researcher in the Geography Research Unit at the University of Oulu, Finland. Her research focuses on sustainability issues, local level tourism development, and the Arctic region.

Marion Karl is a Postdoctoral Researcher at the Department of Geography, LMU Munich, Germany. She concentrates her research on travel decision-making and relevant influencing factors, among them perceptions of risk.

Chase C. Lamborn is a PhD student in the Department of Environment and Society at Utah State University, USA. His research focuses on climate change adaptation amongst fishers, outfitter/guides, the fishing industry, and fisheries management agencies.

Natalie Ooi is Assistant Professor and Program Director of the Ski Area Management Program at Colorado State University, USA. Her areas of research include sustainable tourism, mountain resort development and impacts, social capital, and backpacker tourism.

Joseph O'Leary is Professor in International Tourism at Colorado State University, USA. His research interests include international and domestic travel and recreation behaviour.

Brooke A. Porter is a specialist in the human dimensions of the fisheries and the marine environment. Her work is focused on developing simple and effective development and conservation strategies for coastal communities. She has worked in various capacities with NGOs, international aid agencies, and educational institutions. She currently serves as a scientific adviser to The Coral Triangle Conservancy, an NGO that focuses on reef protection and restoration in the Philippines, and as an Adjunct Professor of Environmental Science at Umbra Institute in Perugia, Italy.

Jarkko Saarinen is Professor of Geography at the University of Oulu, Finland, and a Distinguished Visiting Professor at the School of Tourism and Hospitality, University of Johannesburg, South Africa. His research interests include tourism and development, sustainability management, tourism-community relations, tourism and climate change, and conservation and wilderness studies.

Jürgen Schmude is Professor of Economic Geography and Tourism Research at the Department of Geography, LMU Munich, Germany. The impact of global change on the tourist industry is a core element of his research.

Katrin Magdalena Schwaiger was a PhD student at the Department of Geography, LMU Munich, Germany. Her research focused on tourism in the Caribbean with special focus on hospitality and sustainability.

Jordan W. Smith is the Director of the Institute of Outdoor Recreation and Tourism and an Assistant Professor in the Department of Environment and Society at Utah State University, USA. His research examines how humans make behavioural and planning adaptations in response to rapidly changing environmental conditions.

John Swigart is a Geospatial Analyst at the National Drought Mitigation Center, and writes code for a variety of drought monitoring applications in the United States and internationally. He received his MA degree in Anthropology from the University of Nebraska, Lincoln.

Kaarina Tervo-Kankare is a Postdoctoral Researcher in the Geography Research Unit at the University of Oulu, Finland. Her current research focuses on human/tourism-environment relations, sustainability, nature-based tourism, global environmental change, and rural and wellbeing tourism.

Aaron Weiskittel, PhD, is Professor of Forest Biometrics and Modeling at the University of Maine, USA. He is the Irving Chair of Forest Ecosystem Management and his primary focus is on the development of applied quantitative tools.

Emily Wilkins earned her MS at the University of Maine and is now a PhD student in the Department of Environment and Society at Utah State University, USA.

Sahar Zavareh is a Research Associate at the Department of Geography, LMU Munich, Germany. Her research is focused on long-term disaster recovery of housing, tourism, and insurance.

Tourism in changing natural environments

Natalie Ooi, Esther Duke and Joseph O'Leary

ABSTRACT

Natural environments and the human interactions that occur within, are continuously changing and evolving. However, increasingly, the impacts of climate change, natural and man-made disasters, economic instability, and other macro-environmental factors, have profound implications on local and global economies, fragile ecosystems, and human cultures and livelihoods. In response, tourism within these natural environments is also changing and evolving rapidly in both developed and developing regions. While at times this is spurred by new opportunities, it is often also the result of resource and user-conflicts and changing environmental circumstances. The articles in this special issue examine the numerous ways in which changes in the natural environment impact tourism, and how the tourism industry is responding and adapting to such changes. Detailed case study examination of human interactions within what are often fragile ecosystems can provide us with important insight on social and ecosystem resiliency, innovation and adaptation, and factors that drive tourism success. This was the focus of a session at the inaugural Tourism Naturally Conference, held in Alghero, Italy, 2–5 October 2016, and formed the basis for this collection of articles. From these contributions, what is evident is that while adaptation can be passive in nature, it can, and should ideally be much more proactive, with individuals and organizations seeking improved knowledge and learning. Such actions will contribute to greater resilience within the tourism industry, whether in response to climate change and its subsequent impacts, or an increasing scarcity of the natural resources upon which tourism relies.

摘要

自然环境和人类之间的相互作用是不断变化和演变的。然而，气候变化、自然灾害和人为灾害、经济不稳定以及其他宏观环境因素对当地和全球经济、脆弱的生态系统、人类文化和生计的影响日益严重。作为回应，这些自然环境中的旅游业也在发达和发展中地区迅速变化和发展。虽然有时这是受到新机会的刺激，但也往往是资源和使用者冲突以及环境变化的结果。本期特刊的文章探讨了自然环境对旅游业的影响，以及旅游业是如何应对和适应这些变化的。在通常脆弱的生态系统中，人类相互作用的详细案例研究可以为我们提供关于社会和生态系统恢复力、创新和适应的重要见解，以及帮助我们了解推动旅游业成功的因素。这是2016年10月2日至5日在意大利阿尔盖罗举行的首届旅游自然会议上的会议焦点，成为本期这一系列文章的基础。从这些文章来看，显而易见的是，虽然适应在本质上是被动的，但它可以，而且应该通过个人和组织寻求学习这方面更新的知识而更积极主动。这些行动将有助于旅游业获得更大的恢复力， 无论是应对气候变化及其后续影响，还是应对旅游业所依赖的自然资源日益稀少。

Introduction

We live in a world of constant change and evolution. As humans, the manner in which we interact with the natural world and adapt to such changes has been the subject of much investigation. This focus on human adaptation and resilience has arguably heightened in recent years, as the planet has experienced an increase in extreme weather events, natural disasters, and more broadly, changing weather patterns (Intergovernmental Panel on Climate Change [IPCC], 2014a). Tourism has not been immune to such change, and the increased rate of environmental change and the growing fragility of human–environment interactions have raised questions pertaining to the resilience of tourism within natural environments, its adaptive capacity, and the ways in which both tourism operators and guests have responded to shocks and disturbances.

Defined as 'the long-term capacity of a system to deal with change and continue to develop' (Stockholm Resilience Center, 2015, p. 3), resilience is a measure of the ability of systems to 'absorb changes of state variables, driving variables, and parameters, and still persist' (Holling, 1973, p. 17). This understanding of resilience shifts the management, planning, and policy perspective from controlling change within assumed stable systems, to sustaining and enhancing 'the capacity of social–ecological systems to cope with, adapt to, and shape change' over time (Folke, 2003, p. 227). This is complementary with the definition of a human ecosystem as a '...coherent system of biophysical and social factors capable of adaptation and sustainability over time' (Burch, Machlis, & Force, 2017, p. 12).

In recognizing that tourism systems are constantly in a state of flux as they respond, adapt, and change in response to various shocks and disturbances, a focus on resilience examines the continued ability of a tourism system to use shocks and disturbances as an opportunity to spur innovation and renewal (Stockholm Resilience Center, 2015). Thus, it can be argued that systems do not become resilient in spite of shocks and disturbances, but rather, because of them (Davoudi, 2012). This ability of a system to prepare, manage, and recover from shocks and disturbances, while building and increasing its capacity for learning and adaptation (adaptive capacity) in a way that does not constrain or erode future opportunities, is a central aspect of resilience (Folke, 2003).

Climate change and its impact on tourism

Numerous shocks and disturbances impact the tourism industry and tourist behavior, with varying outcomes. While shocks and disturbances to systems are often acute (e.g. natural disasters), they can also present themselves slowly over time in the form of slow-change variables, with climate change being one such example. Although climate undergoes natural cycles of variability, with its daily manifestation referred to as weather, long-term systemic changes to the earth's climate have brought about warmer temperatures, changes to precipitation patterns, and more frequent and intense weather events (Becken, 2013a). These changes can profoundly affect the economic, social, and environmental systems, upon which tourism relies, making climate change a pressing issue within tourism research (Becken, 2013b).

As a weather-dependent industry, many tourism destinations owe their popularity to their climates (Amelung, Nicholls, & Viner, 2007; Rutty & Scott, 2010). Mountain destinations across the world rely on natural snowfall and cool temperatures to provide downhill skiing and snowboarding and other winter recreation activities (Fukushima, Kureha, Ozaki, Fujimori, & Harasawa, 2002; Hennessy et al., 2008; Moen & Fredman, 2007; Scott, Dawson, & Jones, 2008; Scott, McBoyle, Minogue, & Mills, 2006; Steiger & Stotter, 2013). Yet climate change has already impacted many mountain tourism destinations through the provision of warmer summers and winters, changes in precipitation, and consequently water supply, and increased extreme weather events (Scott et al., 2006).

Coastal tourism and island nations similarly rely on climate to attract tourists around the world looking for pristine beaches and warm sunny weather. However, with changes in climate, many are facing increased vulnerability as coral reefs are bleached by rising sea temperatures (IPCC, 2014b) and islands and atolls face accelerating shorelines and beach erosion from sea-level change (Becken, Hay, & Espiner, 2011; Perch-Nielsen, 2009). The phenomenon of 'last chance tourism' has also arisen and been spurred on by changes in climate from Antarctica to the Maldives to the Great Barrier Reef (Becken et al., 2011; Coghlan, 2013; Lamers, Eijgelaar, & Amelung, 2011). While in the short term, the media attention provided to these 'disappearing' destinations can lead to a surge in tourism, the long-term resilience of these destinations must be brought into question as the ecosystems for which the destination is valued, are irreversibly altered (Coghlan, 2013; Lamers et al., 2011). Furthermore, the paradoxical nature of last chance tourism is such that negative impacts on these environments are further accelerated by increased visitation numbers spurred by this desire for tourists to visit before disappearance (Eijgelaar, Thaper, & Peeters, 2010; Lamers et al., 2011).

The role of climate change in altering tourism visitation and demand to these destinations must therefore be recognized. As the research by Hadwen, Arthington, Boon, Taylor, and Fellows (2011) demonstrates, climate can be a driving force in the seasonal patterns of visitation in certain areas. This is no surprise, given that changes to temperature and precipitation levels can affect the quality of tourism recreation experiences, particularly those that occur outdoors (e.g. national park visitation). This can in turn affect both tourist's willingness to pay, and their likelihood to visit and participate in certain recreational activities (Richardson & Loomis, 2005). While changes in climate in some destinations may bring about positive changes to tourism demand and spending behavior (Richardson & Loomis, 2005), it is likely to bring about adverse effects in others (Scott, Jones, & Konopek, 2007). Either way, there will likely be substantial implications regarding the spatial and temporal redistribution of tourism activities (Amelung et al., 2007).

While these changes to average temperature and precipitation conditions can significantly impact tourism visitation and the desirability of particular destinations, climate change has also brought about increased climate variability and extreme weather events (Van Aalst, 2006), which can subsequently lead to natural disasters, with devastating impacts on tourism. According to IPCC (2014b), as a result of climate change, it is very likely that heat waves will occur more frequently and last longer, and that extreme precipitation events will become more intense and frequent in many regions. This leaves human natural systems exposed to increased risk of hazardous events and trends, such as increased tropical cyclones and droughts, which are contributing to greater economic losses, and devastation to human population and natural areas at rates faster than risk reduction can be achieved (United Nations International Strategy for Disaster Reduction [UNISDR], 2015). This highlights the vulnerability of many natural and human systems, and the tourism industries that operate within.

Tourism as an agent of change

It is also important to note that while tourism around the world relies heavily upon the natural environment within which it operates, it can be the instigator for change, both positive and negative. This is especially evident when impacts at various scales (e.g. local, regional, and global) are considered. For example, air travel from San Francisco, U.S.A. to visit Mount Kenya National Park in Kenya contributes carbon emission which incrementally exacerbates climate change, while at the same time entrance fees paid to visit the park and lodging fee paid to a local ecotourism operator/rancher may contribute to local wildlife management capacity and habitat conservation or even restoration of degraded habitat on private cattle ranches converted to wildlife ranches.

At the national level, especially in many biodiversity-rich countries with developing economies, the revenue generated through tourism provides the much-needed funding for basic conservation activities. For example, iconic protected areas around the world, such as the Monteverde Cloud Forest Reserve in Costa Rica, the Galapagos National Park of Ecuador, and the Volcanoes National Park, in Rwanda, rely on large amounts of tourism income to pay for the cost of running and maintaining these sites (Font, Cochrane & Tapper, 2004). This dependence on tourism is also recognized by Lindsey et al. (2017), who note that a significant determinant of the level of wildlife conservation is the economic value of wildlife to a country. Safari tourism destinations in Africa top the list of high performers with wildlife-based tourism contributing greatly to their national Gross Domestic Product (GDP).

Such tourism interest can help contribute to species conservation, especially given that managers/operators need to conserve wildlife populations in order to deliver the wildlife experiences that people come to these areas to see. As such, wildlife is now regarded as a valuable tourism asset. However, with significant increases in tourism come significant increases in development and infrastructure to support tourists at their expected level of comfort. Often, development may over-burden fragile ecosystems (e.g. commercial and residential water use, roads, crowding). Such changes, whether natural or man-made, can also lead to environmental scarcity and/or degradation, negatively impacting tourism enterprises and/or competing sources of community livelihood. This is often the case when tourism is proposed as an alternative to other natural resource-based industries, such as agriculture, fisheries, timber production, and oil and gas extraction, as well as amenity migration and the accompanying property and commercial real estate development that are common in many resort communities around the world (Ouimet et al., 2017).

A case study approach to examining the role and effects of tourism in changing natural environments

This special issue of *Tourism Geographies* examines the numerous ways in which changes in the natural environment impact tourism, and also how the tourism industry is responding and adapting to such changes. This was the theme of a featured session at the inaugural Tourism Naturally Conference, held in Alghero, Italy, 2–5 October 2016. Many of the papers in this special issue were presented at this conference, with some additional papers added, given their focus on tourism responses to changing natural environments. All of the articles included in this special issue draw on case study examples from around the world, allowing for detailed exploration of current phenomena within the social and environmental context in which they occur, to provide powerful and in-depth insights (Yin, 2003).

Several articles in this special issue explore changes in the natural environment that are occurring as a result of climate change, and what this means for both tourism businesses and tourist behavior. Tervo-Kankare, Kajan, and Saarinen (2018) examine perceived changes in the natural and socioeconomic environments from climate change, and how primary and secondary tourism businesses in two tourism-dependent communities in northern Finland are adapting to these changes. Results indicate that both benefits and costs can be attributed to changing snow cover, with evaluated benefits seemingly exceeding the costs. However, there also appears to be more tendencies towards passive adaptation, with business decisions made based on past experience and occurred changes, as opposed to taking a more proactive approach. While this approach may work for the short term, this paper highlights the need for longer-term strategies and the creation of a more diversified industry to enhance the resilience of the tourism system in this region.

Jedd et al. (2018) focus on the other side of the industry, tourist behavior, and demand, and create a framework that combines temperature and precipitation data to explore the effects of climate change on tourist behavior. In this paper, this framework is applied to summer tourism visitation data at four national parks in the U.S. to examine how tourism visitation responds to temperature and precipitation extremes. Results indicate greater vulnerability for these national parks in extreme dry conditions, as opposed to extreme wet conditions, with decreased visitation resulting in significant declines in tourist-spending in these regions. This framework is a useful tool for tourism providers seeking to adjust and plan for a range of future climatic conditions.

Similarly, Smith, Wilkins, Gayle, and Lamborn (2018) examine how climate variables affect and predict visitation to five southern Utah national parks across a regional tourism system. The results show how variables like precipitation and temperature are not universal, with care needed in making subsequent assumptions. Also, findings indicate that scale is an important consideration in understanding climate and visitation, with differences between climate variables that are regionally and locally significant drivers of tourism visitation.

Wilkins, de Urioste-Stone, Weiskittel, and Gabe (2018) also explore potential changes to tourist flows and expenditure driven by climate change and its associated impacts. Their case study analysis of tourist behavior in Maine, U.S.A., acknowledges the heterogeneity of tourist behavior, and their motivations, values, goals, and perceptions towards climate change. Segmentation analysis identified three segments of Maine tourists, each of which exhibited differences regarding their weather sensitivity, concern for climate change, and behavior intentions to mitigate the effects of climate change. These results show how a changing climate can alter tourism behavior and demand, and what groups may be the most responsive and beneficial to target for marketing or educational efforts to reduce the impact of local climate change.

Natural disasters can also bring about immediate changes to the environment in which tourism operates, with small island states in places like the Caribbean becoming increasingly prone to tropical storms and cyclones. Schmude, Zavareh, Schwaiger, and Karl (2018) take a small regional scale approach to examining the socioeconomic consequences resulting from Tropical Storm Erika on Dominica by applying a Micro-Level Assessment Model (MLAM) for disaster planning. Results demonstrate that for Tropical Storm Erika, the impact on tourism varied, due to geographic location, and recovery efforts were also hampered by poorly developed infrastructure. The application of MLAM to the island of Dominica also highlights that specific regions face severe challenges and the need for tourism operations and community livelihoods to change and adapt, given the increased prevalence of natural disasters. By identifying variations in regional impacts, this research can help ensure that adequate government subsidies are provided in areas of need and future tourism development strategies are better prepared for natural disasters, thereby minimizing future financial and social losses.

While the above articles highlight various ways in which tourism has responded to changing natural environments, some acute, and others occurring over a longer timeframe, it is also important to recognize that tourism is at times responsible for change itself. The introduction of tourism often challenges existing community livelihoods as it competes for access and control over finite and valuable natural resources. Porter (2018) examines tensions that exist between tourism and other sources of community livelihood. Her case study of Hawai'i, highlights conflicts between extractive and non-extractive marine resource use among stakeholder groups from tourism and the ornamental aquarium trade. Through a thematic analysis of public testimony records from a proposed legislation aimed to establish an aquatic life conservation program that would restrict and regulate the collection of aquatic life, what becomes clear is that at the root of this resource-user conflict is differing public opinion as to the health of aquatic reef populations, and therefore the need for conservation and protection. This gives rise to significant challenges for building tourism resiliency, as the future health of marine tourism in this region is arguably dependent upon greater restrictions placed on the ornamental aquarium trade.

Conclusions

Given the changing natural environment in which tourism finds itself situated, and at times finds itself a contributor towards, the adaptive capacity of both tourists and tourism operations is of increasing relevance. This collection of articles provides insight into the role that tourism can play as an agent of environmental change, but also as a somewhat 'unwilling actor' in natural environments that are undergoing broader macro-environmental and socioeconomic change. From these contributions to the research, what is evident is that environmental conditions are in a constant state of flux and therefore require communities and societies to respond by continuously evolving (Berkes, Colding, & Folke, 2000). While adaptation can be passive in nature, it can, and should ideally be much more proactive, with individuals and organizations seeking improved knowledge and learning, whether on potential changes to tourist behavior, or changing environmental circumstances. Such actions will result in greater resilience within the tourism industry, whether in response to climate change and its subsequent impacts, or the increasing conflicts due to scarcity of the natural resources upon which tourism relies.

Disclosure statement

No potential conflict of interest was reported by the authors.

References

Amelung, B., Nicholls, S., & Viner, D. (2007). Implications of global climate change for tourism flows and seasonality. *Journal of Travel Research, 45*(3), 285–296.

Becken, S. (2013a). Developing a framework for assessing resilience of tourism sub-systems to climatic factors. *Annals of Tourism Research, 43*, 506–528.

Becken, S. (2013b). A review of tourism and climate change as an evolving knowledge domain. *Tourism Management Perspectives, 6*, 53–62. doi:10.1016/j.tmp.2012.11.006

Becken, S., Hay, J., & Espiner, S. (2011). The risk of climate change for tourism in the Maldives. In J. Carlsen & R. Butler (Eds.), *Island tourism: Sustainable perspectives* (pp. 72–84). Wallingford, CT: CAB International.

Berkes, F., Colding, J., & Folke, C. (2000). Rediscovery of traditional ecological knowledge as adaptive management. *Ecological Applications, 10*(5), 1251–1262. doi:10.2307/2641280

Burch, W., Jr., Machlis, G., & Force, J. E. (2017). *The structure and dynamics of human ecosystems: Toward a model for understand and action.* New Haven, CT: Yale University Press.

Coghlan, A. (2013). Last chance tourism on the Great Barrier Reef. In. H. Lemelin, J. Dawson, & E. Stewart (Eds.). *Last Chance Tourism: Adapting Tourism Opportunities in a Changing World* (pp. 133–149). Abingdon: Routledge.

Davoudi, S. (2012). Resilience: A bridging concept or a dead end? *Planning Theory & Practice, 13*(2), 299–333.

Eijgelaar, E., Thaper, C., & Peeters, P. (2010). Antarctic cruise tourism: The paradoxes of ambassadorship, "last chance tourism" and greenhouse gas emissions. *Journal of Sustainable Tourism, 18*(3), 337–354.

Folke, C. (2003). Socio-ecological resilience and behavioral responses. In B. Hansson, A. Biel, & M. Martensson (Eds.), *Individual and structural determinants of environmental practice* (pp. 226–242). Burlington, VT: Ashgate.

Font, X., Cochrane, J., & Tapper, R. (2004). *Tourism for protected area financing: Understanding tourism revenues for effective management plans.* Leeds: Leeds Metropolitan University.

Fukushima, T., Kureha, M., Ozaki, N., Fujimori, Y., & Harasawa, H. (2002). Influences of air temperature change on leisure industries: Case study on ski activities. *Mitigation and Adaptation Strategies for Global Change, 7*(2), 173–189.

Hadwen, K. J., Arthington, A. H., Boon, P. I., Taylor, B., & Fellows, C. S. (2011). Do climatic or institutional factors drive seasonal patterns of tourism visitation to protected areas across diverse climate zones in eastern Australia? *Tourism Geographies, 13*(2), 187–208.

Hennessy, K. J., Whetton, P. H., Walsh, K., Smith, I. N., Bathols, J. M., Hutchinson, M., & Sharples, J. (2008). Climate change effects on snow conditions in mainland Australia and adaptation at ski resorts through snowmaking. *Climate Research, 35*(3), 255–270. doi:10.3354/cr00706

Holling, C. S. (1973). Resilience and stability of ecological systems. *Annual Review of Ecology and Systematics, 4*, 1–23. doi:10.2307/2096802

Intergovernmental Panel on Climate Change. (2014a). Climate change 2014: Impacts, adaptation, and vulnerability. Part A: Global and sectoral aspects. Contribution of Working Group II to the fifth assessment report of the intergovernmental panel on climate change. In C. B. Field, V. R. Barros, D. J. Dokken, K. J. Mach, M. D. Mastrandrea, T. E. Bilir, M. Chatterjee, K. L. Ebi, Y. O. Estrada, R. C. Genova, B. Girma, E. S. Kissel, A. N. Levy, S. MacCracken, P. R. Mastrandrea, & L. L. White (Eds.), (p. 1132). Cambridge: Cambridge University Press.

Intergovernmental Panel on Climate Change. (2014b). Climate change 2014: Synthesis report. Contribution of Working Groups I, II and III to the fifth assessment report of the Intergovernmental Panel on Climate Change. In R. K. Pachauri & L. A. Meyer (Eds.), (p. 151). Geneva: IPCC.

Jedd, T. M., Hayes, M. J., Carrillo, C. M., Haigh, T., Chizinski, C. J., & Swigart, J. (2018). Measuring park visitation vulnerability to climate extremes in U.S. Rockies National Park tourism. *Tourism Geographies, 20*(2), 224–249. doi:10.1080/14616688.2017.1377283

Kaján, E., Tervo-Kankare, K., & Saarinen, J. (2014). Cost of adaptation to climate change in tourism: Methodological challenges and trends for future studies in adaptation. *Scandinavian Journal of Hospitality and Tourism, 15*(3), 311–317. doi:10.1080/15022250.2014.970665

Lamers, M. A. J., Eijgelaar, E., & Amelung, B. (2011). Last chance tourism in Antarctica – Cruising for change? In *R.H. Lemelin, J. Dawson & E.J. Stewart. Last-chance tourism: Adapting tourism opportunities in a changing world* (pp. 25–41). Abingdon: Routledge.

Lindsey, P. A., Chapron, G., Petracca, L. S., Burnham, D., Hayward, M. W., Henschel, P., ... Ripple, W. J. (2017). Relative efforts of countries to conserve world's megafauna. *Global Ecology and Conservation, 10*, 243–252.

Moen, J., & Fredman, P. (2007). Effects of climate change on alpine skiing in Sweden. *Journal of Sustainable Tourism, 15*(4), 418–437. doi:10.2167/jost624.0

Ouimet, P., Clark, P., Lai, B., MacMillan, B., McCaul, J., Talley, G., Vallee, P., Viscasillas, A., & Young, J. (2017). *Destination next: A strategic road map for the next generation of global destination organizations.* Washington, D.C.: Destinations International.

Perch-Nielsen, S. (2009). The vulnerability of beach tourism to climate change – An index approach. *Climatic Change, 100*(3/4), 769–815.

Porter, B. A. (2018). Exploring stakeholder groups through a testimony analysis on the Hawaiian aquarium trade. *Tourism Geographies, 20*(2), 308–330. doi:10.1080/14616688.2017.1375971

Richardson, R. B., & Loomis, J. B. (2005). Climate change and recreation benefits in an Alpine National Park. *Journal of Leisure Research, 37*(3), 307–320.

Rutty, M., & Scott, D. (2010). Will the Mediterranean become "too hot" for tourism? A reassessment. *Tourism and Hospitality Planning & Development, 7*(3), 267–281. doi:10.1080/1479053X.2010.502386.

Schmude, J., Zavareh, S., Schwaiger, K. M., & Karl, M. (2018). Micro-level assessment of regional and local disaster impacts in tourist destinations. *Tourism Geographies, 20*(2), 290–308. doi:10.1080/14616688.2018.1438506

Scott, D., Dawson, J., & Jones, B. (2008). Climate change vulnerability of the US Northeast winter recreation-tourism sector. *Mitigation and Adaptation Strategies for Global Change, 13*(5–6), 577–596.

Scott, D., Jones, B., & Konopek, J. (2007). Implications of climate and environmental change for nature-based tourism in the Canadian Rocky Mountains: A case study of Waterton Lakes National Park. *Tourism Management, 28*(2), 570–579.

Scott, D., McBoyle, G., Minogue, A., & Mills, B. (2006). Climate change and the sustainability of ski-based tourism in eastern North America: A reassessment. *Journal of Sustainable Tourism, 14*(4), 376–398.

Smith, J. W., Wilkins, E., Gayle, R., & Lamborn, C. C. (2018). Climate and visitation to Utah's "Mighty 5" National Parks. *Tourism Geographies*, *20*(2), 250–272. doi:10.1080/14616688.2018.1437767

Steiger, R., & Stotter, J. (2013). Climate change impact assessment of ski tourism in Tyrol. *Tourism Geographies, 15*(4), 577–600.

Stockholm Resilience Centre. (2015). *What is resilience? An introduction to social–ecological research.* Stockholm: Stockholm University.

Tervo-Kankare, K., Kajan, E., Saarinen, J. (2018). Costs and benefits of environmental change: Tourism industry's responses in Arctic Finland. *Tourism Geographies*, *20*(2), 202–223. doi:10.1080/14616688.2017.1375973

United Nations International Strategy for Disaster Reduction [UNISDR]. (2015). *Making Development Sustainable: The Future of Disaster Risk Management. Global assessment report on disaster risk reduction.* Geneva, Switzerland: United Nations Office for Disaster Risk Reducation (UNISDR).

Van Aalst, M. K. (2006). The impacts of climate change on the risk of natural disasters. *Disasters, 30*(1), 5–18.

Wilkins, E., de Urioste-Stone, S., Weiskittel, A., & Gabe, T. (2018). Weather sensitivity and climate change perceptions of tourists: A segmentation analysis. *Tourism Geographies, 20*(2), 273–289. doi:10.1080/14616688.2017.1399437

Yin, R. K. (2003). *Applications of case study research* (2nd ed.). Thousand Oaks, CA: SAGE Publications.

Costs and benefits of environmental change: tourism industry's responses in Arctic Finland

Kaarina Tervo-Kankare, Eva Kaján and Jarkko Saarinen

ABSTRACT

Recent research has focused on the impacts of environmental change to tourism. In particular, the perceived costs of climate change have been increasingly studied. However, the relationship between costs and benefits resulting from the changing environmental conditions for the industry has been less examined. This paper identifies the locally observed changes in the natural and socio-economic environments and aims to analyse the financial costs and benefits to tourism businesses in two tourism-dependent communities in northern Finland. The specific focus is on adaptation and adaptive management in a tourist destination scale. Adaption is understood as an investment creating not only implementation costs, but potentially also benefits for tourism operations. Research materials were collected among tourism and tourism-related businesses through 41 semi-structured thematic interviews. Results indicate that the evaluated benefits of environmental change seem to exceed those of costs. This conforms to the on-going discourse of climate change–tourism relations associated with the Arctic region where both awareness and vulnerability to change are considered relatively high but the level of responses, i.e. adaptation, low. These results can help to further identify the most vulnerable sectors in tourism and assist entrepreneurs preparing for environmental and climate change. However, the paper concludes that while global environmental change, with specific adaptive management strategies, may create local short-term direct benefits for the industry, a long-term sustainability of tourism in the Arctic calls for mitigation responses to climate change.

摘要

近期旅游研究关注环境变化对旅游业的冲击，尤其日益关注对气候变化代价的感知，但是对于环境变化引起的旅游业代价与收益之间的关系却鲜有探讨。本文研究了芬兰北部当地居民观察到的自然与社会经济环境的变化，并且分析了环境变化给当地两个旅游社区的旅游企业带来的经济代价与收益。本文特别关注目的地尺度旅游业对环境变化的适应及相关的适应性管理。适应可以理解为一种投资，不仅产生实施的代价，也会产生潜在的旅游运营收益。本文通过对41位旅游及相关企业进行半结构的主题访谈收集研究资料。结果表明，旅游企业评估的环境变化给旅游业带来的收益似乎超过了因之产生的代价。这符合极地地区气候变化与旅游关系持续讨论的结果，那里不管对气候变化的认识还是对气候变化的敏感性水平都比较高，但是相关的响应水平（比如适应）低。该研究结果有助于旅游业进一步识别出最脆弱的部门，有助于企业家应对环境与气候变化。然而，本文得出结论，尽管全球环境变化由于当地特定的适应管理策略短期内给旅游业带来了收益，但是该地区旅游业若实现长期的持续发展需要响应对策缓解气候变化的冲击。

Introduction

The relationship between tourism and environmental change has been studied for a relatively long time (see Mathieson & Wall, 1982). As one of the largest sectors of the global economy (UNWTO, 2015), the tourism industry is a major contributor to environmental change, including socio-cultural and economic issues. Thus, the impacts of tourism activities have formed the core focus of research (Gössling, 2002; Holden, 2006; Scott, Higham, Gössling, & Peeters, 2013). In tourism research, the management of the impacts and governance of tourism development towards sustainability have formed the main aims in recent decades (Saarinen, 2014). At the same time, it has been widely realised that tourism is largely dependent on and affected by environmental conditions (Gössling & Hall, 2006), as well as highly vulnerable to any changes taking place within its physical, socio-cultural or economic environments. This, in principle, makes the need for sustainability management important for the industry itself (Bramwell & Lane, 2011; Hall, 2013) as the on-going global environmental change (GEC) may have serious consequences for the future prospects of tourism.

In recent decades, there has been an increasing interest in the impacts of global climate change (GCC) on tourism (see Gössling & Hall, 2006; Reddy & Wilkes, 2013). Indeed, climate is often seen as one of the most important resources of a tourist destination (Rutty & Scott, 2010). Instead of mitigation, i.e. limiting and regulating the impacts, this interest has mainly focused on adaption perspectives. A Google Scholar search, for example, provides well over twice as many hits for 'tourism and adaptation' than 'tourism and mitigation'. This also reflects the recent focus in the Intergovernmental Panel on Climate Change (IPCC, 2014) assessments and the United Nations Framework Convention on Climate Change (UNCCC) Paris Agreement (UNCCC, 2015) emphasising the urgent need for adaptation and focus on vulnerability issues. Therefore, while mitigation is critical for long-term survival, coping with the estimated, inevitable changes has become crucial in many parts of societies and in different scales ranging from local to transnational governance (see Bramwell & Lane, 2011; Scott et al., 2013).

In tourism studies, adaptation, referring to actions aiming to reduce the negative effects (and to benefit from the positive effects) of climate change (Smit & Wandel, 2006), has been studied for decades (see Kaján & Saarinen, 2013; Scott, Hall, & Gössling, 2012; Wall, Harrison, Kinnaird, McBoyle, & Quinlan, 1986). The studies have often focused on the adaptation perceptions, attitudes and needs of tourism operators (Brouder & Lundmark, 2011; Hall, 2006; Hopkins, 2014; Tervo, 2008; Rauken & Kelman, 2012; Wolfsegger, Gössling, & Scott, 2008), and on the determinants of adaptation at the business' level (Hoffmann, Sprengel, Ziegler, Kolb, & Abegg, 2009). Typically, the scale of analysis has been local and based on a single tourism operator or business sector, such as down-hill skiing (e.g. Dawson & Scott, 2013; Haanpää, Juhola, & Landauer, 2014; Morrison & Picker-ing, 2013). These kinds of studies are important in order to develop a basic understanding of climate change–tourism relations and adaptation needs in the industry.

In this paper, however, environmental change in relation to tourism is approached in the wider destination context. Thus, the research focus includes the core primary tourism businesses, but there are also enterprises, which receive tourism-related income, but for whom tourism may not be a primary, but secondary, source of revenue. They may also serve the primary tourism businesses working directly with tourists' consumption needs. These secondary tourism businesses, such as retail, transportation or janitorial services are economically important and typical for tourism destinations (see Saarinen, 2003). They may have similar or different vulnerabilities to climatic conditions than core tourism oper-ators. Although economically important and highly typical for tourism destinations, hardly any information exists about climate change adaptation among them (see Kaján, Tervo-Kankare, & Saarinen, 2014). Therefore, this article makes an effort to move away from purely sectoral adaptation efforts to a more destination-based approach, which represents a novel approach. This means including all the businesses operating in a tourism destina-tion which have a full or partial, direct or indirect role, in tourists' consumption. Here the 'destination' refers to a relatively coherent spatial unit that includes tourism (primary) and tourism-related (secondary) businesses and other actors collaborating in co-production, local value-chains and/or marketing (see Saarinen, 2004).

Understanding the destination beyond core tourism businesses alone is crucial when considering the cost and/or benefits of GEC in tourism or the future prospects of tourism-dependent communities in general. A wider perspective is needed which includes socio-cultural and economic (and political) environments, the characteristics (such as social capi-tal), and changes of which may affect adaptation and adaptive capacity to environmental changes (Aall, 2012; Adger, 2003; Scott et al., 2012; Tervo-Kankare, 2012). In addition, envi-ronmental change in relation to tourism is approached here from the perspectives of costs and benefits. The combined costs-benefits approach has not been widely applied to tour-ism research, with the majority of existing tourism studies focusing on the cost element alone (e.g. Kaján et al., 2014; Morrison & Pickering, 2013).

Due to the current modes of governance, the focus on businesses and their percep-tions, preferences and decision-making can be seen as increasingly important perspec-tives in GEC/GCC studies. In general, governance represents a new form of public management structured along market (or quasi-market) organisational models (Jessop, 2002; Rhodes, 1996) where governing structures no longer focus primarily on the tradi-tional roles of public sector government (Hall, 2011). Instead, they increasingly incorporate 'a range of interests drawn from the private sector' (Amore & Hall, 2016, p. 2). Thus, deci-sion-making and responsibilities in adaptation (and mitigation) are also devolved to mar-kets and private-sector operators in a local scale (see Nalau, Preston, & Maloney, 2015). As a result there is a strong emphasis that adaptation 'should be decentralised to the lowest level of governance' (Marshall, 2008, p. 80). In general, this governance perspective high-lights the need to understand how businesses operate and perceive the potential, or exist-ing, costs and benefits of GEC. Here the identification of costs and benefits is based on a quantitative approach. Quantitative aims and results may often be more understandable for the tourism operators (mainly small- and medium-sized enterprises, SMEs) and espe-cially for policy-makers (see Ingirige, Jones, & Proverbs, 2008; Veal, 2006) as decision-making processes are easier to complete based on numerical than on qualitative outcomes. However, this approach also involves challenges as many issues and changes related to the operational environment of businesses are difficult to quantify. Moreover, this article analyses tourism stakeholders' understandings about the changes taking place in socio-economic environment and their relationship with the changes in natural environment (see Eriksen, Nightingale, & Eakin, 2015). Therefore, 'environmental change' refers to not only climate change, especially when focusing on costs and benefits, but also to other changes in the operational 'local' environment.

Tourism and environmental change: adaptation perspectives

Growing public awareness and discussions of GCC have created an intensified interest in environmental change and tourism-related research and policy-making (Becken & Clapcott, 2011; Bramwell & Lane, 2008). This interest has mainly focused on adaptation perspectives, and the concept of 'adaptation' is widely discussed in the literature (see Aall & Hoyer, 2005; Adger, Arnell, & Tompkins, 2005; Ford & King, 2015; Füssel & Klein, 2006; Pelling, 2011; Scott & Becken, 2010). The concept focuses on how a unit, or a system, aims to adapt/adapts to change through transforming its operations (Kelly & Adger, 2000; Pielke, 1998), and is usually seen as a local scale response (see Nalau et al., 2015). In addition, however, adaptation and related actions should contribute to climate resilient development which increases the systems' (climate and weather) stress tolerance and the ability to reorganise or continue operations in changing environments (Lew, 2014). In this respect, adaptation is a critical aspect to resilience and together (Espiner, Orchiston, & Higham, 2017) they can enhance sustainable development locally and in the wider tourism system and societies.

On one hand, adaptation generates costs which may in certain cases hinder or even prevent action (see Steiger & Stötter, 2013). In addition to the problem of not knowing the future unit or system adapting to change, the challenge in estimating the costs of change lies in the complexity of the adaptation process (Kaján et al., 2014). It is also sometimes difficult to define what is considered adaptation and which activities are related to, and considered to be part of, the 'normal' product development in tourism businesses. For example, artificial snow-making can represent an adaption mechanism to changing environmental conditions in some locations (Haanpää et al., 2014; Träwöger, 2014). However, many destinations have traditionally used it, and increasingly continue to do so, to extend the skiing season based on normal, i.e. natural winter conditions.

On the other hand, while adaptation represents a cost, it aims to create benefits for the units and systems transforming their activities in order to manage operations and survive changing environment. Therefore, adaptation can be interpreted as an investment where the exact net benefits are difficult to measure. In particular, the benefits of adaptation towards estimated changes in the future are extremely difficult to study due to the conceptual ground of adaptation: as the future characteristics of 'units or systems' adapting to change are not known in present, it is highly hypothetical to evaluate the matter. Therefore, this research limits its focus on the current perceived changes in the environment and adaption towards those changes. In addition to adaptation being beneficial to enterprises (at least in the long run), the environmental changes may similarly bring with them direct benefits. These benefits may be manifested, for example, in the form of competitive advantage over other enterprises and destination regions, or decreasing infrastructural costs (e.g. warming decreasing the use of heating energy) (Saarinen & Tervo, 2006).

The focus on businesses and their perceptions, preferences and decision-making is increasingly important perspective in GEC/GCC studies. This is due to the current modes of neoliberal governance emphasising (or assuming) the key and active role of markets to deal with their environmental consequences and possible negative externalities of economic growth for environment (see Jessop, 2002, Rhodes, 1996). Based on this, governing structures increasingly incorporate certain interests that originate from the private sector (Amore & Hall, 2016; see Hall, 2011). Thus, decision-making and responsibilities in adaptation (and mitigation) are also devolved to markets and private-sector operators, which highlight the issue of adaptive management in businesses. In general, adaptive management studies have focused on resource and species management contexts (e.g. Armitage et al., 2009; Berkes, Colding, & Folke, 2000; White, Cornett, & Wolter, 2017), but it can be applied to businesses and their relations to changing environments.

Adaptive management is called 'adaptive' as it recognises that environmental conditions are in constant change. This requires communities and societies to respond by continuously evolving (Berkes et al., 2000). In general, adaptive management can be organised as either passive or active adaptive management (Table 1), depending on how learning and knowledge creation take place. Passive adaptive management is reactive and involves learning based on experienced situations and only if it improves decision outcomes. In contrast, active adaptive management involves proactive learning and decisions improving knowledge and learning are valued (Walters, 1986). Both approaches involve learning but active adaptive management is more driven by search for knowledge and information before decisions and actions while in passive adaptive management, learning is more based on experience.

Tourism is an industry, where small enterprises dominate the field, and consequently the costs of any 'extra-curricular' activity may turn into an insurmountable obstacle. Therefore, it is important to be able to assess the sources and amount of costs emerging from the consequences of change. In addition, as Smit, Burton, Klein, and Wandel (2000) have stated, adaptation to climate change in tourism is not restricted to adjusting to changes in the long-term mean climate variables only, as also variability, which includes the weather extremes, is changing. Thus, in tourism, where different kinds of climate variability play an important role in the day-to-day operations (Becken, 2012; Rauken & Kelman, 2012; Shih, Nicholls & Holecek, 2008; Tervo, 2008), adaptation is not limited to temperature differences only. Research has shown that snow conditions, occurrence and timing of extremely cold, hot, windy and rainy days, are important factors in nature-based tourism industry (see Csete & Szécsi 2015; Hopkins, 2014; Nicholls, Holecek, & Noh, 2008; Rauken, Kelman, Jacobsen, & Hovelsrud, 2010). These previous studies have also indicated that the

Table 1. Adaptive management approaches (authors' own compilation).

Active adaptation	Passive adaptation
Proactive (incl. learning)	Reactive (incl. learning)
Driven by search for knowledge and information	Experiences and incidents driven
More voluntary – can be/often is guided externally	'Must do' – often forced by external actors (e.g. policies, laws etc.)
Seen as an investment	Seen as a cost

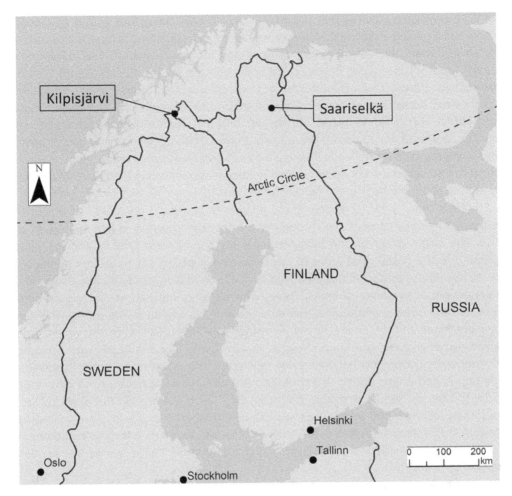

Figure 1. Map of the study area (Authors' own production).

level of climate change adaptation is rather low in many tourism businesses (Cheablam & Shrestha 2015; Haanpää et al., 2014; Hall, 2006; Matasci, Kruse, Barawid, & Thalmann, 2014; Tervo-Kankare, 2011; Träwöger, 2014). Consequently, detailed information about the costs or benefits of (non-existing) adaptation is difficult to obtain. When only little information is available, temporal and spatial analogues could help in developing a basic understanding of the future conditions and the development needs of nature-based industries (Ford et al., 2010).

Study design, methods and data

The study was conducted in Saariselkä (located in Inari municipality) and Kilpisjärvi (located in Enontekiö municipality) in late 2013 and early 2014. Both destinations are located in Finnish Lapland and can be considered peripheral and Arctic tourism destinations located in high latitudes (Figure 1). Even though tourism in both destinations is, to some extent, year-round tourism, they have high seasons: winter in Saariselkä and summer in Kilpisjärvi (Laatutiimi, 2011; Enontekiön matkailualueen turvallisuussuunnitelma,

2013). However, the activities they provide for tourists have similarities, with the exception of Saariselkä having a down-hill skiing centre. In both destinations, nature-based tourism has an important role as a provider of work and income to people and communities nearby: tourism has enabled development in the villages, for example, by bringing in and maintaining services that would not be possible on the basis of permanent population, and they are dependent on tourism income (Kaján, 2014; Saarinen, 2003). Both destinations attract visitors from Finland and abroad, and the origin of tourists affects the seasons. In Saariselkä, for example, the tourists from Asia characterise early winter season (and also summertime) while the Russians arrive in great numbers in early January. Finnish tourists' main seasons occur during the ski-holiday season in spring (February–March) and during the autumn foliage in the fall (Laatutiimi, 2011).

Empirical data were collected through semi-structured interviews, which lasted between 10 and 60 minutes and through an Internet-based survey in late 2013 and early 2014. The target group for the interviews and the survey were local businesses, which operated in the postal code areas of Saariselkä and Kilpisjärvi. In both communities, all businesses whose contact information was available were contacted and requested for an interview either by phone or email. Those, who were not reached at the original stage, were sent the questions through either email or with a link to an online survey. The interview questions and the online-survey were identical. They dealt with respondents' perceived experiences concerning changes in economic, socio-cultural and ecological environments during the last five years, the costs and benefits occurring because of weather-related events during that time slot, and with the basic background information of the businesses (season, activities, revenues, etc.). The questions did not explicitly focus on climate change, but on weather variations and phenomena and their subsequent impacts. This approach is often used in climate change studies as it is more comprehensible and less value-loaded for participants than the notion of climate change (see e.g. Saarinen & Tervo, 2006).

In Saariselkä, 18 businesses were interviewed in person, and five businesses participated via the online version of the survey. Altogether, these 23 businesses represent 33 per cent of the local businesses registered in Saariselkä, where 69 businesses were operating in 2011 (Statistics Finland, 2014). According to the statistical database (Statistics Finland, 2014), Kilpisjärvi has 21 registered businesses of which 15 were interviewed face-to-face and three via email ($n = 18$). This represents 86 per cent of all the businesses in Kilpisjärvi.

Both communities are located in rural Finland and are small in size. Their special characteristics include strong entrepreneurial activity, and a dependency on tourism and nature as a livelihood (Kaján, 2014). All interviews were tape-recorded (when allowed by the interviewees) and the data were compiled and partly transcribed to an Excel file for further processing. The collected data were analysed via qualitative content analysis, where responses were studied in order to find similarities, common or shared views and themes (e.g. phenomena that could be related to climate change on the basis of scientific knowledge about the topic), and linkages between them. Classification and quantification of the responses was utilised when possible. As the sample size is rather small, no statistical analyses were conducted – only certain classifications and distributions were examined using statistical methods such as cross-tabulation and non-parametric tests.

Table 2. The respondents by sector and their turnover figures by village.

Number of respondents	Kilpisjärvi 18	Saariselkä 23	Total 41
Sector			
Accommodation and restaurant services	5	7	12
Retail trade	2	5	7
Programme and recreation services	1	5	6
Transport and repairs	1	0	1
Combinations	3	5	8
Other	6	1	7
Turnover			
<10,000 €	2	1	3
10,000–49,999 €	3	1	4
50,000–99,999 €	2	1	3
100,000–499,999 €*	1	6	7
500,000–999,999 €	3	2	5
≥1,000,000 €	3	8	11
Not known	4	4	8

*Turnover 'hundreds of thousands' placed in this category.

Overall, the number of interviewed businesses (*n* = 41) gives a 47 per cent response rate among all businesses in Kilpisjärvi and Saariselkä. The majority of the participating businesses were from the 'accommodation and restaurant sector', followed by 'combination businesses' (enterprises that provide services falling under two sectors) and 'retail and programme services' (see Table 2). The enterprises were classified on the basis of their dominating activity if the business had one or two clear ones. Thus, the group 'Other' consists of businesses that either offered a multitude (more than two) of equally important activities, or activities that did not fall in the main categories (e.g. translation, janitorial services). Over 75 per cent of all the businesses were willing, and able, to give their annual turnover figures. In addition, the turnover figures for three businesses were available from business-related open databases. As the aim of the paper is to compare the risen expenses and/or financial benefits occurred by different weather phenomena against the annual turnover, the partially limited availability of turnover figures has, to some extent, restricted the analysis.

The study utilises a temporal analogues approach, which is a commonly used method in climate change studies (Ford et al., 2010; Rutty et al., 2017), including tourism research.

Table 3. The perceived changes in natural environment by category (number of remarks if >1).

Season changes				
Snow cover changes (amount of snow, arrival and melting of snow) (*n* = 24)	Warming of winters (*n* = 14)	Warming of summers (*n* = 13)	Timing and length of seasons (winter, spring, summer) (*n* = 9)	Coldness of summers (*n* = 3)
Weather changes				
Increase in occurrence of extreme weather (*n* = 4)	Increasing wind strength (*n* = 3)	Less predictability of weather	Less extreme weather phenomena	
Flora and fauna				
Occurrence of lemmings, large autumnal carpet (a moth) (*n* = 4)	Not enough lichen for reindeer (too many reindeer in the area)	Tree growth	Number of squirrels has increased	Changes in the timing and size of mushroom crops
Other				
Ash cloud (*n* = 2)	Mistiness because of snow-making	Increasing construction may threat landscapes	Water shortage	Mining boom

Both Dubois and Ceron (2006) and Dawson, Scott, and McBoyle (2009) have utilised and examined the usability of analogues in tourism adaptation studies. Temporal analogues, in comparison to basic climate change future scenarios, allow the consideration of how human systems operate 'within a framework of coexisting stresses' (Rutty et al., 2017, p. 197), i.e. the human behaviour and reactions to diverse climate phenomena. Therefore, analogues also consider the socio-economic environment that affect and guide human behaviour. This approach is important in studies aiming to understand the areas of extra costs and benefits: sectors, seasons, and types of costs and benefits (labour-related, pro- grammes, infrastructure, sales).

Changing environment and adaptation in Kilpisjärvi and Saariselkä

Perceived changes in natural environment

In the two destinations, the majority of the respondents (88 per cent) had experienced changes in their natural environment. Mostly, the observed changes related to issues that originated from changes in the climatic conditions (Table 3). They can be categorised as changes taking place in relation to seasons, weather conditions in general, flora and fauna, and other. The 'other' category includes separate or sporadic matters such as water short- ages. To some extent, the season- and weather-related changes are interrelated; therefore this classification should not be interpreted as categorical. In addition, as humans often experience 'weather' rather than 'climate', the categorisation is somewhat approximate rather than precise interpretation of the experienced changes. Geographically, the notions about changes did not vary considerably.

Though sometimes the observed phenomena were conflicting, the most common issues referred to changes in snow cover. What is interesting is the diversity of observed phenomena within a relatively short time period (five years, compared to the timeframe associated with climate change in general). The respondents did not always attach posi- tive or negative significance to the observed changes in the context of their business. Therefore, the changes cannot be categorised according to their relevance in terms of infrastructure, sales, labour or programme management. Instead, the changes were men- tioned in relation to their personal lives.

Perceived changes in socio-economic environment

As environmental changes do not exist in an isolated vacuum, the interviews also gath- ered information about locally experienced changes outside the climate change narrative and natural settings alone. Out of all respondents, 88 per cent had observed changes in their socio-economic environment during the last five years. Four thematic groups emerged from the material: infrastructure, economic issues, sales and customers. When compared to changes in natural environment, the respondents were more often able to pinpoint the positive and negative impacts of these changes.

The positive infrastructural changes included increased diversification of tourism serv- ices and the renewed network of skiing and hiking routes in both locations. Also, increased permanent housing was seen as positive as it contributes to the destination's stability. Despite the increase in construction, the built environment was still seen as spacious and well-built in Saariselkä. The main negative point related to infrastructure in Saariselkä concerned the local airport services and flight connections, which were considered unfair in comparison to competitive destinations (expensive connections) and also unreliable to some extent (mainly due to airline strikes). Other negative aspects were linked with power issues and dominance among the accommodation establishments. Certain businesses were considered to have power over the others and it was noted in both destinations, that there was a certain social distance between the newcomers and more permanent residents. It was also stated that the hotel chains were not so committed to local tourism development and had other values than local establishments. Electricity power cuts were an issue in both communities. Both locations also perceived a decrease in tourism services. Paradoxically in Kilpisjärvi this was partially the result of an outmigration of population and not so much due to the declining attractiveness as a tourism destination as is often the reason for decreasing services.

The global financial crisis that started in 2007, and the recession that followed till the fieldwork period in 2014, were quoted to be noticeable factors in terms of both local and global economics and resulted in less sales and difficulties in planning. Also changes in taxation were a negative factor affecting the socio-economic development in both locations. Growth factors were only observed in Saariselkä, whereas in Kilpisjärvi, it was stated that the unknown status of the potential new national park in the immediate area was stalling further investments (the Finnish government decided against establishing the national park due to local resistance in the spring of 2014 after fieldwork).

The positive sales and customer-related factors in Kilpisjärvi were associated with a more diverse customer base through an increase in Russians and independent tourists. Also, Saariselkä had observed an increase in visitation from Russian markets, and foreign markets were stated not to be affected by the recession and their numbers had in fact increased. Also, the amount of Norwegians coming for holidays and taking advantage of trans-boundary shopping opportunities increased. This was coupled with an increase in Finnish tourists. Despite the diversified customer base, the changes in tourists' typology were also considered as negative socio-economic development, as both destinations were experiencing losses in their main segments. The main segment has traditionally been the independent cross-country skiers, but this group of tourists is getting older and hence smaller. The activity needs of the younger generation are different resulting in the permanent decrease in the previously known main segment.

Concerns were related to the high dependency on the Norwegian market and the currency fluctuations concerning Norwegian crown (krone). It was also remarked that even though Norwegians were a very dominant group, the economic benefits did not disperse well throughout the community. The increased amount of tourists interested in snowmobiling was seen as both beneficial and distracting as it was creating conflicts between un motorised and motorised programmes. The spring tourism season was perceived to be shorter and creating pressure on other seasons. In Saariselkä, it was stated that the combination of airline strikes, recession and poor sales were detrimental to the destination. Tourism development was also affected by increased international competition. The current trend also indicated an increase in small, rather than large, groups.

Table 4. Benefits and their significance according to sector, category, amount of benefit, reason and location (S = Saariselkä, K = Kilpisjärvi) (benefit = % of annual turnover).

Sector	Category	Benefit = % of annual turnover	Reason	Location
Experienced phenomena				
Accommodation and restaurant	Sales	30.7%	Sunny days in spring	K
Accommodation and restaurant	Sales	0.1%	Heat in the south	K
Transport	Sales	0.5%	Bad weather	K
Combination (accommodation and restaurant services; Recreation and programme services)	Sales	6.3%	No snow in the south	K
Emerging phenomena				
Accommodation and restaurant	Sales, infrastructure	2.0%	Great amount of snow + warm winter	K
Recreation and programme services	Sales	0.1%	Less extreme cold days	S
Retail	Sales	0.02%	Great amount of snow	K
Combination (accommodation and restaurant; transport)	Sales	0.1%	Warmer, longer summer	K
Other (janitorial services)	Sales	0.2%	More snow	K
Other (journalist)	Sales	33.3%	Exceptional weather conditions	K

Benefits of environmental change

Altogether, 78 per cent of the respondents indicated that they had benefitted financially from the exceptional weather conditions taking place during their main season. Almost half of them (47 per cent; which is 37 per cent of all respondents) were able to give a numerical estimation of the benefit. In general, the financial benefits were rather insignificant, mostly varying between 0.1 and 2 per cent of the annual turnover (see Table 4). Occasionally, the benefits were much higher, reaching 33 per cent. In all but one case, the end result of the benefits was manifested as increased *sales*. *Infrastructural* benefits were mentioned once, in relation to decreasing energy demand. *Programme management*- or *labour*-wise no direct benefits occurred.

In general, the positive impacts were related to two broad themes. The first is related to *experienced* conditions, which mainly consisted of local weather phenomena happening under normal variability (mentioned 17 times). Also, weather and climatic phenomena or changes taking place in visitors' home regions (mentioned four times) fall under this category. These phenomena were considered as 'normal' and did not raise much discussion. On the contrary, the second theme is related to *emerging* phenomena, where local climatic conditions can be associated with climate change (mentioned 16 times) referring to more permanent changes.

The experienced phenomena mainly referred to the extremes of 'good' and 'bad' weather, and to the quick weather fluctuations. Sunny weather increased the outdoor sales whereas rain and storms attracted consumers to inside activities. In general, the benefits occurred at the expense of someone else's business. Estimating the significance of the benefits under this category seems to be difficult: only in two out of 17 cases (12 per cent) the interviewees were able to give numerical value to the benefit (in relation to annual turnover). In addition, the weather and climatic phenomena taking place in visitors' home regions were mainly related with the climatic counterpoints in tourism: people want to travel to north to escape the heat or to see snow in occasions, when no snow is present at home. In two out of four cases, the respondents gave a numerical value for the benefit.

The emerging phenomena referred to the changing amounts of snow and warming and the increasing number of unusual weather-related phenomena. This kind of phenomena can be related to climate change, as they are, according to climate science, signs of GCC in the study area. Mostly, the benefits regarding the amount of snow referred to the great amount of snow, which causes more work and results in more income. It also increases customer satisfaction and makes the destinations more attractive through relative snow security. Moreover, the respondents were able to estimate the benefits under this category as 44 per cent provided a numerical estimate for the benefit. Other benefits from the emerging phenomena included increased indoors sales due to bad weather (e.g. stormy/rainy weather that may prevent outdoor activities) and lower energy costs due to warming.

In terms of GEC-related benefits per sector, the results are the following in a descending order: other (average benefits 16.8 per cent), accommodation and restaurant (10.9 per cent), combination enterprises (3.2 per cent), transport (0.5 per cent), recreation and programme services (0.1 per cent) and retail (0.02 per cent). To conclude, 5 out of 6 (83 per cent) recreation and programme service businesses had benefitted from climate and weather phenomena, while in accommodation and restaurant services 8 out of 12 (67 per cent) entrepreneurs reported benefits. Under the categories retail and transport services, all of the businesses (100 per cent) had received extra income due to climatic and weather conditions. Also in the other and combination sectors, the majority of businesses (71 and 75 per cent, respectively) had experienced benefits. The average benefit created by changing environment, including climate change, among the businesses was 7.3 per cent in relation to annual turnover.

Costs of environmental change

The root-causes for extra costs were connected to bad weather, unreliable weather forecasts, snow-related issues, warm weather, rain and wind (see Table 5). These weather-

Table 5. Costs generated by different weather elements according to sector, category, amount of cost, reason and location (S = Saariselkä, K = Kilpisjärvi).

Sector	Category	Cost = % of annual turnover	Reason for loss	Location
Accommodation and restaurant	Sales, infrastructure	0.8%	Bad weather	K
Recreation and programme services	Sales	17.5%	Rain and winds	K
Recreation and programme services	Programme management, sales	1%	Consistency of snow	S
Recreation and programme services	Infrastructure	0.3%	Lack of snow/too much ice	S
Retail	Sales	20%	Bad fall weather	S
Retail	Sales, infrastructure	0.9%	Excessive amount of snow	K
Combination (recreation and programme services; accommodation and restaurant)	Labour	2.0%	Unpredictable weather	S
Combination (recreation and programme services; Accommodation and restaurant)	Labour and sales	0.1%	Long fall	S

related extra expenses can be divided into four categories: *sales, infrastructure, tourism programmes* and *labour costs*. Though the sample size is fairly small, there is indication that the actual costs accrued and the financial impacts of environmental change are still relatively insignificant among many tourism service providers. However, of all the respondents (*n* = 41), 44 per cent (= 18 enterprises) had experienced extra expenses generated by climatic shifts. Almost half (44 per cent) of them were able to give cost estimations. Thus, 19.5 per cent of all the respondents estimated the amount of costs. Most commonly, the extra costs were generated by the decreased sales figures.

Though the estimated costs are seemingly low figures, it must be remembered that entrepreneurs in Finnish Lapland are mostly small businesses that may suffer significantly financially from one single weather event. For example, exceptionally warm fall weather delayed the winter sales reducing as much as 20 per cent of overall sales with one company, and rain and wind had reduced sales for almost 18 per cent with another activity provider in a recent year. This can be considered a significant financial impact. The greatest individual loss (in absolute terms) was generated by a weather incident, namely blizzard, when one of the main roads was blocked by snow, which prevented customers' access, and led to loss in sales. However, this particular loss in combination with other weather-related costs represents merely 0.9 per cent of this company's entire annual turnover.

In another company, costs emerged from the company's own vulnerable infrastructure and decreased sales. These impacts were caused by bad weather and frequent power cuts that caused harm to electronic equipment and caused cancellations, forcing them to reduce prices. The costs represented roughly 0.8 per cent of the annual turnover, whereas benefits were far more, approximately 30 per cent. Other infrastructure related costs were mostly based on the increased snow-ploughing due to excessive snow or on the impacts of power cuts due to storms. One respondent indicated that the snowstorm had caused 8000€ additional costs in snow-ploughing but in relation to turnover, the economic impacts were insignificant (0.3 per cent).

The labour costs were linked with having too much workforce due to poor sales and/or having to employ new staff due to disproportionately high snowfall. In both of these occasions, the source for labour costs related with snowfall. Additional labour costs also occurred when weather forecasts failed and the weather prevented the planned operations or maintenance work. In one business, the additional labour-related costs were quoted to be several tens of thousands, but in relation to turnover, this was merely 0.1 per cent. In contrast, the daily sales could increase by 100 per cent in nice weather for the same company. The extra costs in safari programmes were generated by re-arranging activities, changing seasons and being extra cautious with programme timings. Often, the costs were related to snow. Other costs concerning the safari programme management were fairly small, less than or around one per cent, and were an inconvenience rather than a significant financial cost.

To conclude, in two occasions the negative impacts generated approximately 18–20 per cent extra costs, which were both related to decrease in direct sales. For the remaining businesses, the increased costs were minor (app. 0.1–1 per cent). The average cost of adaptation among the respondents who reported and evaluated the amount of costs (*n* = 8) was 5.1 per cent. When calculating by sector, the average costs were the following:

retail 10.5 per cent, recreation and programmes 6.3 per cent, accommodation and restaurants 0.8 per cent, and combination establishments 0.1 per cent.

Sector-wise, 50 per cent of recreation and programme service enterprises reported costs from climate and weather phenomena, while the same issue was reported in the accommodation and restaurant services by 42 per cent of the entrepreneurs. In retail services, over half of the enterprises (57 per cent) had suffered from extra costs because of climatic and weather conditions. In the sector other, none of the seven businesses reported extra costs, and in combination businesses the costs had occurred among 63 per cent of the respondents. In transport-category, the only representative (100 per cent) reported costs.

Discussion

This paper aimed to evaluate the costs and benefits of environmental change and related (potential) adaptation measures to businesses operating in tourism destinations in two Arctic communities in Finnish Lapland. While focusing on adaptation, the issue of GEC and its local scale impacts were highlighted with respect to the evaluation of costs and benefits. The benefits of changing environmental (e.g. weather) conditions were concretised via changing consumption patterns and increased sales while costs were directed towards a quite specific part of business operations (e.g. certain element of infrastructure, work force and salary costs).

As the sample size was limited, analyses to identify statistically reliable differences were also limited. In addition, the turnover data covered only 80 per cent of the enterprises participating in the study, which decreased the accuracy of calculations. However, some comparisons concerning the amount of extra costs and benefits and their appearance were run on the basis of sector, size of the business (annual turnover), location and the main season. Of the respondents, 44 per cent stated to have experienced weather-related costs and the numerical responses specified that the average cost of adaptation was 5.1 per cent. In return, the experienced benefits affected 78 per cent of the entrepreneurs resulting in an average of 7.3 per cent increased turnover. Socio-economic changes were related to infrastructure, economic issues, sales, and customer-base, and were found to affect tourism development by 88 per cent of the interviewees. The most commonly referred changes in the natural environment were related to seasons, weather conditions, and flora and fauna. These were observed by the majority, namely 88 per cent of the respondents.

Concerning the types of adaptation, the results show more tendencies for passive adaptation, where the decisions are made based on experiences and occurred changes. As shown above, 88 per cent of the interviewees have experienced changes in their natural environment, with sentiments varying between negative and positive impacts. The experienced changes and the subsequent benefits that were felt by 78 per cent of the respondents, emphasise passive adaptation. According to the both types of adaptation (reactive and passive) there is an element of learning present, which is also detectible in the results. For example the changing shoulder seasons are nowadays to some extent approached with caution. Shoulder seasons are understood as the time between high and low seasons, more specifically relating to late autumn and late spring as well as early summer. Furthermore, though snow security exists, the experience has shown that also the conditions in tourist originating regions impact the destinations.

Despite the prominence of passive adaptation, understanding *emerging* climatic conditions shows an element of active adaptation. Therefore, though the costs are mostly related to passive adaption there is an occasional active adaptation investment present. Occasionally the costs occur due to just struggling with the general environmental change and variability. The results signal a relatively low financial impact of environmental change related issues on tourism entrepreneurs in Lapland. However, the real costs may be far greater and the abstract notion of climate change must be taken into consideration as people are not always able to synthesise between local level issues and climate change impacts. In this sense, perhaps the changes are 'not yet serious enough'.

Also, adaptation mechanisms, and their meanings, vary according to destinations. For example, the snow-making was not considered to be an adaptation action by locals but rather a normal activity related to the unpredictability of the climatic conditions. Therefore, no specific sum was given to this activity. In tourism adaptation research, however, snow-making is regarded as one of the main adaptive mechanisms in snow-based tourism (see Haanpää et al., 2014; Träwöger, 2014).

It is crucial to remind that this paper has evaluated the costs and benefits of GEC in a tourism destination context, which leaves many larger scale and indirect impacts unconsidered. Some references to these kinds of impacts were made, such as the reliance on airport services and the municipality/regional level infrastructure: for example, the costs that took place when the community's infrastructure failed. The interviewed entrepreneurs were also more often able to estimate the net effect of phenomena that can be associated with climate change than phenomena that falls in category 'normal variability'. This may indicate two things: first, the phenomena relating to climate change are new and therefore registered better, and second, the net effect of normal variability evens out in time when both beneficial and detrimental climatic events fluctuate.

When identifying the combined effects of costs and benefits per sector, certain trends arise. Accommodation and restaurant sector seems to be benefitting the most from the changes in weather fluctuations. Retail can be assessed to suffer the most along with recreation and programmes whereas transport and other sectors are only benefitting as they did not report any costs. Moreover, the combination enterprises also seem to mainly receive benefits. This implies that diversification is an effective strategy against climate variability. Also, in relation to geographical location, Kilpisjärvi seems to benefit more than Saariselkä. Season-wise, the highest benefits take place in businesses with snow season as their main season. This contradicts to some extent with several studies which indicate vulnerability of snow-dependent tourism and resilience of the so-called summer tourism in Northern Europe (Amelung, Nicholls, & Viner, 2007; Grillakis, Koutroulis, Seiradakis, & Tsanis, 2016; Saarinen & Tervo, 2006). The studies show that volatile snow-conditions create unpredictability for businesses where the changing winter conditions are difficult to manage. Substituting activities in winter is far more challenging than in summer as the difference between not having snow, and having snow, is crucial (Agnew & Viner, 2001; Saarinen & Tervo, 2006).

Interestingly, certain phenomena may mean high costs for some sectors while others seem to benefit from their occurrence, reducing the destination-wide costs. For example, bad weather was stated to benefit transport sector while in retail, accommodation and restaurant, recreation and programme services this caused costs. The same applied to vast amount of snow: it benefitted accommodation and restaurant services and janitorial services while recreation and programme services suffered. In retail sector, both benefits and costs were registered for the same reason.

Therefore, it is important to understand that the net benefits/costs of adaptation to environmental changes may depend considerably on the scale of analysis: individual business may suffer even though at the destination scale net benefits are recorded. Another interesting factor about the combination and other groups is that it could indicate emerging new professions or give information about emerging new benefitting sectors, which do not fit in the traditional categories. In addition, it is important to examine the relative occurrence of both costs and benefits sector-wise: the most unaffected sector in terms of costs seems to be the other sector while in relation to benefits retail seems to be receiving the most benefits.

There is some indication that in terms of costs, the establishments with high turnovers ($\geq 1,000,000€$) were able to absorb the costs more efficiently. However, as not all businesses were comfortable with sharing information concerning their turnover, these results are only indicative. Bearing these facts in mind, it can be stated that of the interviewed business that were able to pinpoint costs, half had turnovers in the highest category with costs adding up to 0.1–0.9 per cent. However, spatial dimension of costs and benefits could add to new knowledge about environmental change and climate change. In this study, however, there were no major differences in the occurrence or amount of costs between the two study destinations even though Kilpisjärvi seemed to benefit more than Saariselkä. Furthermore, comparing the net effects of environmental change in different regions/destinations and even at different times could specify the most economically vulnerable regions and seasons. For example, a study focusing on eastern Lapland by Kaján et al. (2014) indicated that the costs of the main peak season (Christmas) adapting to climate change in relation to annual turnover are similar to this study (between 1 and 5 per cent), but that the costs in relation to seasonal turnover were as high as 250 per cent. In the southern tourist destinations in Finland, adaptation to winter conditions can be expected to be even more costly (Saarinen & Tervo, 2006) but the financial benefits are still largely unknown.

If there are alternatives replacing the outdoor activities during bad or unpredictable weather, the overall tourist experience is not necessarily negative (see also Denstadli, Jacobsen, & Lohmann, 2011). More diversified services can create more benefits to the entire community (Kaján, 2014). As the shoulder seasons are expected to lengthen, summers to suffer from more precipitation and the snow-conditions to change (see IPCC, 2014), the role of diversified industry plays a key role. However, in order to build sustainable and climate-resilient destinations, both the costs and benefits should be equally divided so that one sector or single enterprise does not take the full burden of costs, nor gain the full benefits at the expense of others. Again, understanding the importance of the scale of analysis is emphasised.

The advantage of establishing the current net adaptation costs lies in their seemingly simple approach, which could be communicated to tourism entrepreneurs relatively easily. This means that the quantified approach may be more easily understood. The used time span of five years is not necessarily long enough in a scientific sense for evaluating the impacts of environmental change, or especially GCC, but in the context of tourism industry, shorter time spans are often needed due to the nature of the businesses and its planning and management cycles (see Lépy et al., 2014; Saarinen & Tervo, 2006). However, in addition to short-term adaptation measures, a neglected focus on mitigation in tourism operations should be emphasised in the future in both tourism systems and at the destination scale. According to Stern (2006), for example, the cost of mitigation is only one per cent of the GDP, while the cost of non-mitigation might be as high as 20 per cent of the GDP. Thus, while adaptation is perhaps perceived more acute for the successful management of tourism operations in changing environmental conditions, the most cost-effective way of reacting to the challenge of climate change is a pro-active approach focusing on mitigation. Obviously, the time span of the benefits of mitigation measures is much longer and focusing on too distant future for SME tourism businesses compared with the adaptation as an investment with faster potential returns.

This refers to the basic challenge in climate change adaptation and mitigation planning: the future units, systems, changes and capacities to respond are not really known, especially when the tourism industry is as dynamic as it is. This relates a so-called time horizon problem, which refers to the tendencies of people and organisations to focus on near, rather than medium and long term futures (see Orlove, 2010). Human actors place low value and, thus, action preference to issues that take place in the distant future. The time horizon problem may also manifest itself as an unsustainable adaptation choice and/ or the lack of interest in mitigation. This makes it challenging to implement the known or estimated future costs to present day decisions and practices (e.g. in tourism-climate change relations) (Saarinen, 2014). As the study has shown, evaluating and understanding the impacts of climate change in different time and spatial scales is not only pertinent but also challenging. In addition, it is important to acknowledge the intertwined nature of any changes in societies and international relations, due to which the drawing of direct conclusions is somewhat difficult.

Conclusions

Current emphasis on neoliberal governance highlights the need to understand how businesses operate, adapt and perceive the potential or existing costs and benefits of GEC. Adaptation is necessary as climate change has advanced, or is about to reach, the stage where adaptation is already required. The results of this study indicate that the current benefits of environmental change and adaptation exceed the costs in local scale. The currently occurring weather phenomena generate more financial income for the local SMEs than produce additional costs. In addition, the results show particular sectoral vulnerabilities. However, the net benefits are attained by passive, rather than active, adaptation efforts. This knowledge may assist in developing more resilient destinations and businesses in and outside tourism. The results also reveal that there may be other benefitting sectors outside the tourism sphere, which role may increase as climate change proceeds.

Overall, however, these results indicate a potential challenge in the current modes of governance emphasising a more market-driven approaches in climate change responses mainly based on a local scale responsibilities. As the environmental change in the Arctic is generally seen by the businesses as beneficial and adaptation measures as a profitable investments for the operations, what would then be the incentives for the industry to be pro-active and, thus, more responsible in the mitigation? The adaptation actions are also hindered by the relatively short planning cycles among SMEs. All this raises critical questions against the current emphasis on decentralisation of adaptation to the lowest level of governance and seeing adaptation mainly as a local scale responsibility, as local level often has very limited capacity and resources to plan and implement for (long-term) adaptation.

Though this paper has focused on measures that mainly deal with adaptation, most probably only the combined efforts of effective adaptation and mitigation can safeguard more sustainable futures for the Arctic communities living and working in fragile environments that are expected to be greatly impacted by the GCC. For this, a more regulative and long-term mode of governance by the public sector and/or public-private partnerships is likely needed to guide the tourism industry to respond locally to the challenge of GEC. Understandings of impacts of GEC are influenced by varying perceptions of change itself, differing time and spatial scales, unpredictable business management issues and estimations of advantages and disadvantages of change.

Disclosure statement

No potential conflict of interest was reported by the authors.

Funding

Academy of Finland [FICCA: The CLICHE-project (Impacts of Climate Change on Arctic Environment, Ecosystem Services and Society)].

References

Aall, C. (2012). The early experiences of local climate change adaptation in Norwegian compared with that of local environmental policy, Local Agenda 21 and local climate change mitigation. *Local Environment, 17*, 579–595. doi:10.1080/13549839.2011.631992

Aall, C., & Hoyer, K. G. (2005). Tourism and climate change adaptation: The Norwegian case. In C. M. Hall & J. Higham (Eds.), *Tourism, recreation and climate change* (pp. 209–221). Clevedon: Channel View Publications.

Adger, N. (2003). Social capital, collective action, and adaptation to climate change. *Economic Geography, 79*(4), 387–404. doi:10.1111/j.1944-8287.2003.tb00220.x

Adger, N., Arnell, N., & Tompkins, E. (2005). Successful adaptation to climate change across scales. *Global Environmental Change, 15*, 77–86. doi:10.1016/j.gloenvcha.2004.12.005

Agnew, M. D., & Viner, D. (2001). Potential impacts of climate change on international tourism. *Tourism and Hospitality Research, 3*(1), 37–60. Retrieved from http://www.jstor.org/stable/23743849

Amelung, B., Nicholls, S., & Viner, D. (2007). Implications of global climate change for tourism flows and seasonality. *Journal of Travel Research, 45*, 285–296. doi:10.1177/0047287506295937

Amore, A., & Hall, C. M. (2016). From governance to meta-governance in tourism? Re-incorporating politics, interests and values in the analysis of tourism governance. *Tourism Recreation Research, 41*(2), 109–122. doi:10.1080/02508281.2016.1151162

Armitage, D. R., Plummer, R., Berkes, F., Arthur, R. I., Charles, A. T., Davidson-Hunt, I. J., ... Wollenberg, E. K. (2009). Adaptive co–management for social–ecological complexity. *Frontiers in Ecology and the Environment, 7*(2), 95–102. doi:10.1890/070089

Becken, S. (2012). Measuring the effect of weather on tourism: A destination- and activity-based analysis. *Journal of Travel Research, 52*(2), 156–167. doi:10.1177/0047287512461569

Becken, S., & Clapcott, R. (2011). National tourism policy for climate change. *Journal of Policy Research in Tourism, Leisure and Events, 3*(1), 1–17. doi:10.1080/19407963.2011.539378

Berkes, F., Colding, J., & Folke, C. (2000). Rediscovery of traditional ecological knowledge as adaptive management. *Ecological Applications, 10*(5), 1251–1262. doi:10.2307/2641280

Bramwell, B., & Lane, B. (2008). Priorities in sustainable tourism research. Journal of Sustainable Tourism, 16(1), 1–4, doi:10.2167/09669580803489612.

Bramwell, B., & Lane, B. (2011). Critical research on the governance of tourism and sustainability. *Journal of Sustainable Tourism, 19*(4–5), 411–421. doi:10.1080/09669582.2011.580586

Brouder, P., & Lundmark, L. (2011). Climate change in Northern Sweden: Intra-regional perceptions of vulnerability among winter-oriented tourism businesses. *Journal of Sustainable Tourism, 19*(8), 919–933. doi:10.1080/09669582.2011.573073

Cheablam, O., & Shrestha, R. P. (2015). Climate change trends and its impact on tourism resources in Mu Ko Surin Marine National Park, Thailand. *Asia Pacific Journal of Tourism Research, 20*(4), 435–454. doi:10.1080/10941665.2014.904803

Csete, M., & Szécsi, N. (2015). The role of tourism management in adaptation to climate change – A study of a European inland area with a diversified tourism supply. *Journal of Sustainable Tourism, 23*(3), 477–496. doi:10.1080/09669582.2014.969735

Dawson, J., & Scott, D. (2013). Managing for climate change in the alpine ski sector. *Tourism Management, 35*, 244–254. doi: 10.1016/j.tourman.2012.07.009

Dawson, J., Scott, D., & McBoyle, G. (2009). Climate change analogue analysis of ski tourism in the northeastern USA. *Climate Research, 39*, 1–9. doi:10.3354/cr00793

Denstadli, J. M., Jacobsen, J. K. S., & Lohmann, M. (2011). Tourist perceptions of summer weather in Scandinavia. *Annals of Tourism Research, 38*, 920–940. doi:10.1016/j.annals.2011.01.005

Dubois, G., & Ceron, J.-P. (2006). Tourism and climate change: Proposals for a research agenda. *Journal of Sustainable Tourism, 14*(4), 399–415. doi:10.2167/jost539.0

Enontekiön matkailualueen turvallisuussuunnitelma. (2013). *(Tourism security plan for Enontekiö region). Matkailualan tutkimus - ja koulutusinstituutti.* Retrieved from http://www.enontekio.fi/media/enonteki-f6n_matkailualueen_turvallisuussuunnitelma_11.3.2013.pdf

Eriksen, S. H., Nightingale, A. J., & Eakin, H. (2015). Reframing adaptation: The political nature of climate change adaptation. *Global Environmental Change, 35*, 523–533. doi:10.1016/j.gloenvcha.2015.09.014

Espiner, S., Orchiston, C., & Higham, J. (2017). Resilience and sustainability: A complementary relationship? Towards a practical conceptual model for the sustainability–resilience nexus in tourism. *Journal of Sustainable Tourism, 25*, 1385–1400. doi:10.1080/09669582.2017.1281929

Ford, J. D., & King, D. (2015). A framework for examining adaptation readiness. *Mitigation and Adaptation Strategies for Global Change, 20*(4), 505–526. doi:10.1007/s11027-013-9505-8

Ford, J. D., Keskitalo, E. H. C., Smith, T., Pearce, T., Berran-Ford, L., Duerden, F., & Smit, B. (2010). Case study and analogue methodologies in climate change vulnerability research. *Wiley Interdisciplinary Reviews, 1*(3), 374–392. doi:10.1002/wcc.48

Füssel, H.-M., & Klein, R. J. T. (2006). Climate change vulnerability assessments: An evolution of conceptual thinking. *Climatic Change, 75*, 301–329. doi: 10.1007/s10584-006-0329-3

Grillakis, M. G., Koutroulis, A. G., Seiradakis, K. D., & Tsanis, I. K. (2016). Implications of 2°C global warming in European summer tourism. *Climate Services, 1*, 30–38. doi:10.1016/j.cliser.2016.01.002

Gössling, S. (2002). Global environmental consequences of tourism. *Global Environmental Change, 12*, 283–302. doi:10.1016/S0959-3780(02)00044-4

Gössling, S., & Hall, C. M. (Eds.). (2006). *Tourism and global environmental change: Ecological, social, economic and politic interrelationships*. London: Routledge.

Haanpää, S., Juhola, S., & Landauer, M. (2014). Adapting to climate change: Perceptions of vulnerability of down-hill ski area operators in Southern and Middle Finland. *Current Issues in Tourism, 18* (10), 966–978. doi:10.1080/13683500.2014.892917

Hall, C. M. (2006). New Zealand tourism entrepreneur attitudes and behaviours with respect to climate change adaptation and mitigation. *International Journal of Innovation and Sustainable Development, 1*(3), 229–237. doi:10.1504/IJISD.2006.012424

Hall, C. M. (2011). A typology of governance and its implications for tourism policy analysis. *Journal of Sustainable Tourism, 19*(4–5), 437–457. doi:10.1080/09669582.2011.570346

Hall, C. M. (2013). Framing behavioural approaches to understand and governing sustainable tourism consumption. *Journal of Sustainable Tourism, 21*(7), 1091–1109. doi:10.1080/09669582.2013.815764

Hoffmann, V. H., Sprengel, D. C., Ziegler, A., Kolb, M., & Abegg, B. (2009). Determinants of corporate adaptation to climate change in winter tourism: An econometric analysis. *Global Environmental Change, 19*(2), 256–264. doi:10.1016/j.gloenvcha.2008.12.002

Holden, A. (2006). *Environment and tourism*. London: Routledge.

Hopkins, D. (2014). The sustainability of climate change adaptation strategies in New Zealand's ski industry: A range of stakeholder perceptions. *Journal of Sustainable Tourism, 22*(1), 107–126. doi:10.1080/09669582.2013.804830

IPCC. (2014). *Climate change 2014: Impacts, adaptation, and vulnerability. Part A: Global and sectoral aspects. Contribution of Working Group II to the FifthAssessment Report of the Intergovernmental Panel on Climate Change*. Cambridge & New York, NY: Cambridge University Press. Retrieved from http://www.ipcc.ch/pdf/assessment-report/ar5/wg2/WGIIAR5-PartA_FINAL.pdf

Ingirige, M. J. B., Jones, K., & Proverbs, D. (2008). *Investigating SME resilience and their adaptive capacities to extreme weather events: A literature review and synthesis*. In R. Haigh & D. Amaratunga (Eds.), *Conference proceedings of the CIB international conference on building education and research: Building resilience (BEAR) 2008, Kandalama, Sri Lanka, 11th–15th February 2008*, (pp. 582–593). Salford: University of Salford. Retrieved from http://www.irbnet.de/daten/iconda/CIB11298.pdf

Jessop, B. (2002). Liberalism, neoliberalism, and urban governance: A state-theoretical perspective. *Antipode, 34*(3), 452–472. doi:10.1111/1467-8330.00250

Kaján, E. (2014). Arctic tourism and sustainable adaptation: Community perspectives to vulnerability and climate change. *Scandinavian Journal of Hospitality and Tourism, 14*(1), 60–79. doi:10.1080/15022250.2014.886097

Kaján, E., & Saarinen, J. (2013). Tourism, climate change and adaptation: A review. *Current Issues in Tourism, 16*(2), 167–195. doi:10.1080/13683500.2013.774323

Kaján, E., Tervo-Kankare, K., & Saarinen, J. (2014). Cost of adaptation to climate change in tourism: Methodological challenges and trends for future studies in adaptation. *Scandinavian Journal of Hospitality and Tourism, 15*(3), 311–317. doi:10.1080/15022250.2014.970665

Kelly, P. M., & Adger, W. N. (2000). Theory and practice in assessing vulnerability to climate change and facilitating adaptation. *Climatic Change, 47*(4), 325–352. doi:10.1023/A:1005627828199

Laatutiimi. (2011). *Saariselän alueen laatukäsikirja* [A handbook of quality for Saariselkä region]. Retrieved from http://www.saariselka.fi/sisalto/laatu/laatu

Lew, A. A. (2014). Scale, change and resilience in community tourism planning. *Tourism Geographies, 16*, 14–22. doi:10.1080/14616688.2013.864325

Lépy, É., Heikkinen, H. I., Karjalainen, T. P., Tervo-Kankare, K., Kauppila, P., Suopajärvi, T., ... Rautio, A. (2014). Multidisciplinary and participatory approach for assessing local vulnerability of tourism industry to climate change. *Scandinavian Journal of Hospitality and Tourism, 14*(1), 41–59. doi:10.1080/15022250.2014.886373

Marshall, G. (2008). Nesting, subsidiarity, and community-based environmental governance beyond the local scale. *International Journal of the Commons, 2*(1), 75–97. doi:10.18352/ijc.50

Matasci, C., Kruse, S., Barawid, N., & Thalmann, P. (2014). Exploring barriers to climate change adaptation in the Swiss tourism sector. *Mitigation and Adaptation Strategies for Global Change, 19*(8), 1239–1254. doi:10.1007/s11027-013-9471-1

Mathieson, A., & Wall, G. (1982). *Tourism: Economic, physical and social impacts.* London: Longman.

Morrison, C., & Pickering, C. M. (2013). Perceptions of climate change impacts, adaptation and limits to adaption in the Australian Alps: The ski-tourism industry and key stakeholders. *Journal of Sustainable Tourism, 21*(2), 173–191. doi:10.1080/09669582.2012.681789

Nalau, J., Preston, B. L., & Maloney, M. C. (2015). Is adaptation a local responsibility? *Environmental Science & Policy, 48,* 89–98. doi:10.1016/j.envsci.2014.12.011

Nicholls, S., Holecek, D. F., & Noh, J. (2008). Impact of weather variability on golfing activity and implications of climate change. *Tourism Analysis, 13*(2), 117–130. doi:10.3727/108354208785664256

Orlove, B. (2010). Time horizons and climate change. *Weather, Climate, and Society, 2*(1), 5–7. doi:10.1175/2009WCAS1111.1

Pelling, M. (2011). *Adaptation to climate change: From resilience to transformation.* London: Routledge.

Pielke, R. A. (1998). Rethinking the role of adaptation in climate policy. *Global Environmental Change, 8*(2), 159–170. doi:10.1016/S0959-3780(98)00011-9

Rauken, T., & Kelman, I. (2012). The indirect influence of weather and climate change on tourism businesses in Northern Norway. *Scandinavian Journal of Hospitality and Tourism, 12*(3), 197–214. doi:10.1080/15022250.2012.724924

Rauken, T., Kelman, I., Jacobsen, J. K. S., & Hovelsrud, G. K. (2010). Who can stop the rain? Perceptions of summer weather effects among small tourism businesses. *Anatolia, 21*(2), 289–304. doi:10.1080/13032917.2010.9687104

Reddy, M. V., & Wilkes, K. (Eds.). (2013). *Tourism, climate change and sustainability.* Oxon: Routledge.

Rhodes, R. (1996). The new governance: Governing without government. *Political Studies, 44*(4), 652–667. doi:10.1111/j.1467-9248.1996.tb01747.x

Rutty, M., & Scott, D. (2010). Will the Mediterranean become "too hot" for tourism? A reassessment. *Tourism and Hospitality Planning & Development, 7*(3), 267–281. doi:10.1080/1479053X.2010.502386

Rutty, M., Scott, D., Johnson, P., Pons, M., Steiger, R., & Vilella, M. (2017). Using ski industry response to climatic variability to assess climate change risk: An analogue study in Eastern Canada. *Tourism Management, 58,* 196–204. doi:10.1016/j.tourman.2016.10.020

Saarinen, J. (2003). The regional economics of tourism in Northern Finland: The socio-economic implications of recent tourism development and future possibilities for regional development. *Scandinavian Journal of Hospitality and Tourism, 3*(2), 91–113. doi:10.1080/15022250310001927

Saarinen, J. (2004). 'Destinations in change': The transformation process of tourist destinations. *Tourist Studies, 4,* 161–179. doi:10.1177/1468797604054381

Saarinen, J. (2014). Critical sustainability: Setting the limits to growth and responsibility in tourism. *Sustainability, 6*(11), 1–17. doi:10.3390/su6010001

Saarinen, J., & Tervo, K. (2006). Perceptions and adaptation strategies of the tourism industry to climate change: The case of Finnish nature-based tourism entrepreneurs. *International Journal of Innovation and Sustainable Development, 1*(3), 214–228. doi:10.1504/IJISD.2006.012423

Scott, D., & Becken, S. (2010). Adapting to climate change and climate policy: Progress, problems and potentials. *Journal of Sustainable Tourism, 18*(3), 283–295. doi:10.1080/09669581003668540

Scott, D., Hall, C. M., & Gössling, S. (2012). *Tourism and climate change: Impacts, adaptation & mitigation.* London: Routledge.

Scott A., Higham, J., Gössling, S., & Peeters P. (Eds.). (2013). *Understanding and governing sustainable tourism mobility: Psychological and behavioural approaches.* London: Routledge.

Shih, C., Nicholls, S., & Holecek, D. F. (2008). Impact of weather on downhill ski lift ticket sales. *Journal of Travel Research, 47*(3), 359–372. doi:10.1177/0047287508321207

Smit, B., & Wandel, J. (2006). Adaptation, adaptive capacity and vulnerability. *Global Environmental Change, 16,* 282–292. doi:10.1016/j.gloenvcha.2006.03.008

Smit, B., Burton, I., Klein, R. J., & Wandel, J. (2000). An anatomy of adaptation to climate change and variability. *Climatic Change, 45*(1), 223–251. doi:10.1023/A:1005661622966

Statistics Finland. (2014). *Enterprises: Establishments by postal code*. 31.12.2013. Retrieved from http://pxnet2.stat.fi/PXWeb/pxweb/en/StatFin/

Steiger, R., & Stötter, J. (2013). Climate change impact assessment of ski tourism in Tyrol. *Tourism Geographies, 15*(4), 577–600. doi:10.1080/14616688.2012.762539

Stern, N. H. (2006). *Stern review: The economics of climate change*. London: HM Treasury.

Tervo, K. (2008). The operational and regional vulnerability of winter tourism to climate variability and change: The case of the Finnish nature–based tourism entrepreneurs. *Scandinavian Journal of Hospitality and Tourism, 8*(4), 317–332. doi:10.1080/15022250802553696

Tervo-Kankare, K. (2011). The consideration of climate change at the tourism destination level in Finland: Coordinated collaboration or talk about weather? *Tourism Planning & Development, 8*(4), 399–414. doi:10.1080/21568316.2011.598180

Tervo-Kankare, K. (2012). Climate change awareness and adaptation in nature-based winter tourism. Regional and operational vulnerabilities in Finland. *Nordia Geographical Publications, 41*(2).

Trawöger, L. (2014). Convinced, ambivalent or annoyed: Tyrolean ski tourism stakeholders and their perceptions of climate change. *Tourism Management, 40*, 338–351. doi:10.1016/j.tourman.2013.07.010

UNFCCC. (2015). *The Paris agreement*. United Nations. Retrieved from http://unfccc.int/files/meetings/paris_nov_2015/application/pdf/paris_agreement_english_pdf

UNWTO. (2015). *UNWTO tourism highlights 2015 edition*. Madrid: World Tourism Organization. Retrieved from http://www.e-unwto.org/doi/pdf/10.18111/9789284416899

Veal, A. J. (2006). *Research methods for leisure and tourism: A practical guide*. Harlow: Pearson Education.

Wall, G., Harrison, R., Kinnaird, V., McBoyle, G., & Quinlan, C. (1986). The implications of climate change for camping in Ontario. *Journal Recreation Research Review, 13*, 50–60.

Walters, C. J. (1986). *Adaptive management of renewable resources*. New York: Mc Graw Hill.

White, M. A., Cornett, M. W., & Wolter, P. T. (2017). Two scales are better than one: Monitoring multiple-use northern temperate forests. *Forest Ecology and Management, 384*, 44–53. doi:10.1016/j.foreco.2016.10.032

Wolfsegger, C., Gössling, S., & Scott, D. (2008). Climate change risk appraisal in the Austrian ski industry. *Tourism Review International, 12*(1), 13–23. doi:10.3727/154427208785899948

Measuring park visitation vulnerability to climate extremes in U.S. Rockies National Parks tourism

Theresa M. Jedd, Michael J. Hayes, Carlos M. Carrillo, Tonya Haigh, Christopher J. Chizinski and John Swigart

ABSTRACT

Changes in temperature and precipitation can affect tourist experiences. This study examines how summer park visitation has changed in response to temperature and precipitation extremes. The study goals were two-fold. The first is to introduce a framework and the second is to test it in a pilot region with four mountainous National Parks. The framework is designed to compare the vulnerability of seasonal park visitation to shifts in a combined indicator of temperature and precipitation. It uniquely considers needed measurements, and the data required to conduct an analysis. The second goal is to test it in four destinations in the U.S. Northern Rockies, including Glacier, Yellowstone, Grand Teton, and Rocky Mountain National Parks. The preliminary test reveals outlier cases of visitation under wet and dry extremes. The analysis connects time series climate and visitation data for the peak summer season from 1991–2012. Outlier analysis illustrates more change in extremely dry conditions, with four out of the six dry-year outliers resulting in a visitation decline. Whether this decline in park tourism is attributable to climate features, economic factors, or conscious management decisions, these drops have significant economic impacts: estimates of changes in visitor spending during dry years are between roughly 9 and 90 million USD. These differences may be connected to the popular activities in each park, and the extent they are dependent on weather conditions. This framework can be used to test the relationship between climate and tourism visitation in other regions, in various seasons and time frames. The work may inform the tourist sector in adjusting and planning for a range of conditions. We discuss opportunities and conclude with additional needs for understanding the mechanisms behind risk in mountain park tourism under climate extremes.

摘要

由于国家公园地区温度与降雨的变化影响旅游体验，所以使国家公园易于受到访问量降低的影响。本研究研究了国家公园夏季访问量面对温度与降雨极端情况所经历的变化。研究目标有二：第一是引入一个框架，该框架用来比较面对温度与降雨复合指标变化国家公园夏季访问量受影响的程度。该框架独特地考虑了所需的测量指标以及执行分析的数据可得性。第二是在有四个国家公园（包括冰川国家公园、黄石国家公园、大蒂顿国家公园和洛基山国家公园）的美国落基山脉北部地区检验了这个框架。初步结果揭示了国家公园访问量在极端降雨与干旱情况下访问量奇高与奇低的极端情况。分析结合了1991–2012年间国家公园夏季旺季气候与访问量的时间序列数据。异常值分析表明，极端干旱天气将导致公园访问量更大的变化，其中6 个极端干旱年份有4个年份导致了访问量的下降。不管公园游客量的降低是由于气候因素经济因素还是人为的管理决策，游客量的降低都产生显著的经济影响：干旱年份游客消费变化的估计值大约在9 百至9 千万美元。游客消费的变化可能与公园主要的旅游活动有关，相关程度取决于天气情况。如果获得有关数据，该框架可以用来检验不同地区、不同季节、不同时间范围内气候与旅游访问量的关系。该项工作可以帮助旅游部门调整与规划一系列变化情况。最后我们讨论了为理解山地公园在极端天气情况下山地旅游风险机制的机遇与额外的需要。

1. Background: tourism in mountain parks

Mountainous regions offer a scenic backdrop for adventurous travel, and outdoor enthusiasts have long sought them out as a destination. National parks in the Northern United States' Rocky Mountains are loved and revered for their scenic beauty (Quammen, 2016). Because of the management protections that govern their use, mountain parks are also recognized as internationally significant study sites (Knowles & Colwell, 2012). These mountainous areas are home to fragile alpine ecosystems that can be detrimentally affected by human settlement and development. Around the world, mountainous areas are often under park-protected status to limit these effects by knitting competing uses and various cultural observances under a unified conservation ethic (Chape et al., 2008). Protected status in mountain areas has been shown to facilitate wildlife movement through the landscape (Goetz et al., 2009) and connects managers with knowledge networks for dealing with common threats like invasive plant species (McDougall et al., 2011). Land managers in the Northern Rockies have relied on park designation as a management tool since the inception of Yellowstone as the first U.S. National Park. Large landscape-scale governance arrangements continue to rely on the National Park Service (NPS) as a key ally in achieving conservation goals.

By necessity, studies of mountain tourism should include measures of park usage (Nepal, 2002). The interplay between tourism, park protection and ecological integrity has been a popular topic of study. Work to date has shown bidirectional effects: park visitation can change in response to observed and projected environmental conditions (Scott, Jones, & Konopek, 2007), and also has an effect on the status and quality of the environment (Monz et al., 2010). Increased visitor loading can lead to trail widening and vegetation trampling (Dale & Weaver, 1974). As tourists move rocks and pick flowers (Willard, 1970), they also modify ecological integrity. Studies to understand park visitation, therefore, have also considered the implications of increased trail use (Loomis & Keske, 2009). These effects are significant, and a deeper consideration of mountain tourism should encompass the drivers behind park visitation. Stevens et al. (2014) consider changes in per capita national park visitation in response to economic variables such as per-vehicle entrance fees, fuel prices, and median incomes, concluding that when travel costs outpace income growth, visits to national parks may decline when they are no longer affordable.

Beyond the realm of economic decision-making, climate conditions may also drive park visitation. This study investigates how mountain park tourism connects with climate conditions, to what extent it is vulnerable to change under extremes, and the possible economic impacts of these changes. It constitutes an initial response to the request for work on tourism in the context of drought vulnerability in Thomas et al. (2013). It also answers the call to understand the dynamic between environmental conditions and tourist behavior (Butler, 2000) by targeting the questions: can yearly fluctuations in visitation be explained by variation in the climate? On another level, how have these visitor fluctuations been connected to climate extremes? Then, if these effects are known, what economic impact might they have and what can the parks do to address these vulnerabilities in the future? To answer these questions, we developed a framework to investigate the link between physical and socio-economic indicators of current and future climate impacts, incorporating scientific and sectoral input to construct an acceptable set of time series proxies for exposure, sensitivity and vulnerability under different climate patterns. The analysis is an answer to the call by the Cooperative and Joint Venture Agreement of the Great Plains Cooperative Ecosystems Study Unit, which states that scientific research should incorporate the physical and social sciences (U.S. Fish and Wildlife Service, 2015).

1.1 *Weather and climatic effects on alpine tourism*

Anecdotal evidence suggests that temperature and precipitation impact the quality of recreational opportunities, but the question remains whether these climate features have a measurable quantifiable effect on the overall amount of tourist visits. Previous studies have examined the effect of temperature on park visitation, showing that visitation generally increases in U.S. parks with temperature until it reaches a threshold of 25 °C, or 77 °F, at which point it declines (Fisichelli, Schuurman, Monahan, & Ziesler, 2015). In warm climates, it is uncomfortable and risky to hike or engage in strenuous physical exertion in hot conditions. Managers recognize the hazard that high temperatures pose to tourists and can use trail closure status and other measures to keep hikers from excessive exposure in hot dry areas with little or no water and sustenance (Proctor pers. comm., 2017).

However, in cooler-climate regions, warming temperatures can have a positive effect, drawing visitors out in larger numbers to experience pleasant conditions. For example, in Canadian National Parks, warming trends were shown to have a positive effect on visitation in the summer as well as the 'shoulder season' spring and fall months (Jones & Scott, 2006). However, warming trends can cause ecological degradation, and intensify outbreaks of the mountain pine beetle (Morris & Walls, 2009), which visitors view as negative and unacceptable (McFarlane & Witson, 2008). Thus, temperature changes may have mixed effects on park tourists in cooler alpine settings.

Under future predictions of warming, precipitation may be the mitigating factor on visitor comfort and ecological function. As with temperature, it is conceivable that tourism in mountainous areas can also shift in either direction as a result of changes in precipitation. For example, dryness can be a benefit in the case of vehicle and foot travel. With decreased snow depth, Rocky Mountain National Park's Trail Ridge Road could be accessible to vehicle and cycle travelers, or hiking trails may be accessible for more of the year (Richardson & Loomis, 2005). Wet weather in parks may worry managers who are concerned with tourist safety (Martín, 2005), and this ranks especially high when considering extreme events like disease outbreaks or flooding linked to climate variability internationally (Agnew & Viner, 2001). Flooding can wash away roads and infrastructure in and around parks. In mountain areas where road networks are likely limited, even one eroded or collapsed section can preclude access to the entire park when it occurs on the major access highway. This dynamic played out at the individual park level in September 2013 when Rocky Mountain experienced major visitor decline and a federally declared disaster for the surrounding areas (Gochis et al., 2015) as the result of a large rain event and damaging floods.

Warmer and drier conditions, on the other hand, can contribute to diminished outdoor opportunities as they are associated with detrimental fires (Westerling et al., 2011), decreased reservoir levels (Pielke et al., 2005), and diminished river flows (Shrestha & Schoengold, 2008). Megafire events can change viewscapes; necessitate trail, road and campground closures; and have devastating effects on the wildlife and vegetation that visitors seek (Delgado pers. comm., 2016). Damaged forest cover and understory vegetation affected ungulate habitat and grazing behavior in Northern Yellowstone in the years following a large fire year in 1988 (Pearson et al., 1995). During this fire season, over 250,000 hectares burned due to unusually severe drought conditions; this was surprising to researchers and managers, as fires had previously been contained by boundaries set by forest age and type (Turner et al., 1994). Though park visitation dropped 15% from the prior year, it climbed to a record high in 1989, indicating to the park community that 'the drop in tourism revenues, like the decline in greenery, was only temporary' (Franke, 2000, p. 3). Parsing out the effects of short-term conditions from longer trends is useful in understanding the ways that weather and climate affect tourism.

Weather is an instantaneous representation of atmospheric conditions; it affects momentary decisions. Weather stations measure wind, temperature, and precipitation (features that people notice when they are outside). Climate is a longer-term aggregation of these values. It is defined by the statistical pattern of the daily weather features over a defined time span (Shea, 2016). The two concepts are related in a nested fashion, with weather fitting inside climate, as it is a momentary snapshot of longer-term averages. Temperature and humidity play into perceptions of how appealing it is to be outside and if/when people choose to engage in particular activities. These short-term choices are made on an ongoing basis and may be adjusted quickly. Climate, on the other hand, has the potential to determine what activities are feasible in an area, or what a location becomes known for. It is unlikely that swimming pools or sunbathing would be popularly marketed vacation activities in cool-climate destination, just as is it equally bizarre to see snow skiing in desert climates.

Similarly, recreation and tourism are often used interchangeably, and are sometimes also discussed as nested concepts. To date, tourism has been defined as 'the practice of travelling for recreation' while recreation is 'an activity in which individuals voluntarily engage for personal satisfaction or pleasure' (de Freitas, 2003). Recreation is rooted in a particular place and time, and can happen close to home, even on a casual basis when neighborhood friends meet to play a game of basketball. However, tourism incorporates a broader sense of geography, in that it is inherently built upon the notion of travel (Hall & Page, 1999; Williams, 1998). Not all tourism is done with the goal of seeking outdoor recreation, though. For example, a visitor to the Czech Republic may stop after a work meeting to visit a cathedral in Prague, or a traveler to London may only be interested in seeing the Big Ben clock tower. However, in the Rocky Mountain alpine park environment, natural and minimally built-up spaces are the norm, placing outdoor recreation as the most visible dimension of tourism.

As such, it follows that weather and climate are central determinants of park-based tourism in these areas. One study on tourist comfort argues the critical importance of considering precipitation as a factor in park travel experiences (de Freitas, Scott, & McBoyle, 2008); other studies have considered temperature trends in tourists visits of parks (Jones & Scott, 2006; Richardson & Loomis, 2004; Scott et al., 2007). Fluctuations can have direct and indirect effects, as they affect comfort and satisfaction in the short term, and perceptions and reputation in the longer term. Short term direct impacts like lower reservoir levels (Jiang et al., 2015) have cascading effects such as increased algal bloom, which can affect the quality of swimming waters. Instream flow reduction affects boaters' and anglers' ability to find white water rapids or fishing holes (Loomis, 2008; Shrestha & Schoengold, 2008), and lower snowpack affects skier enjoyment and wildflower seeker success in high alpine environments (Lambert, Miller-Rushing, & Inouye, 2010; Pederson et al., 2011).

Media-based studies suggest reduced precipitation can lead to fishing tournament cancellation, boat ramp extension, boat damage by exposed debris or stumps in shallow areas, fishing camp decline, firework bans, golf course irrigation, and hunter's reduced access to wildlife (Dow, 2010). Over time, secondary, or indirect impacts, stem from bad publicity and/or negative perceptions: in the case of drought, these include decreased visitations, cancellations in hotel stays, or a reduction in booked holidays (Thomas et al., 2013). Secondary impacts are also caused by decreased overall tourist satisfaction when changes in temperature and precipitation begin to affect the ecology and wildlife of an area.

Not all tourist experiences are equally affected, and in some cases this means that certain opportunities will be diminished while others flourish. In the Western United States, mountainous areas known for snow readjust to decreased snowpack and rely more on hiking and mountain biking than on skiing. For example, one high-elevation community has rebranded itself to include other activities implied with the name 'Telluride Ski & Golf,' while Purgatory Resort in Durango, Colorado, has added a zipline, alpine slide, and ropes course. This type of activity substitution plays a role in adaptation (Klein, 2011).

2. Conceptualizing vulnerability

At its heart, a vulnerability study begins with an examination of the extent to which individuals and communities are likely to be harmed as a result of the physical impacts of a natural hazard or threat (Frazier, 1979), and builds up to the social and political dimensions of the systems designed to cope with them. As such, it is an endeavor that spans many disciplines. More than the threat posed by a natural hazard, it is a relative consideration of social justice, equity, and available opportunities (Eakin & Luers, 2006). It has been described as individuals' and groups' ability to *cope with* external stresses placed on livelihoods and well-being (emphasis added, Adger & Kelly, 1999). It also comprises the characteristics that enable a group to *resist and recover* from the impacts of natural hazards (emphasis added, Blaikie, et al., 1994).

A vulnerability study considers a complement of factors that either make an activity susceptible or resilient to changes (González Tánago et al., 2016). Building a framework involves defining key concepts and making choices about which measurements or data sources will be used to connect the pieces of information (Ostrom, 2009). Risk-hazard research offers an outcome-based approach and measures of the economic and social losses associated with environmental change (Hayes, Wilhelmi, & Knutson, 2004). It views outcomes as determined by a chain of events beginning with the stressor, which leads into its impacts, and is followed by the adjustment/adaptation response. As such, outcome variable change can answer the following questions. What is the sector vulnerable to? What consequences can be expected? And when or where would these impacts of exposure to the initial stressor occur? This can take the form of the conceptual equation, which is informed by the Intergovernmental Panel on Climate Change (2014) perspective:

$$\text{Vulnerability} = \frac{\text{sentitivity} \times \text{exposure}}{\text{adaptive capacity}} \tag{1}$$

Physical measures of vulnerability are based on functional relationships between particular biophysical stressors and the resulting damages (e.g. changes in water availability and income). In pairing social and physical dimensions, indicators should be chosen to match societal concerns. For tourism, a sector that depends on public perceptions, it is particularly important to consider which outcomes matter most to the community.[1] Thomas et al. (2013) identified hotel stays, visitation numbers, sales tax revenues, fuel volumes, visitor spending, and employment numbers as important markers. In another study, de Freitas (2003) examined economic decision making from a hypothetical tourist willingness-to-pay approach; in this view, experiences matter to the extent that tourists pay for them.[2,3] As an outcome variable, park visitation numbers are a commonly used indicator for tourism levels (Eagles, 2014). Because these numbers fluctuate, they can be used to establish the magnitude and direction of deviation from the average for the amount of recreation in a park.

2.1 Exposure to physical changes

Proper identification of exposure is a multi-step process: one needs to determine the metrics associated with a hazard, and the length of time that is required to feel its effects (days, months, or years). Climate data is used as a diagnostic tool to identify deviations from normal conditions. Studies have done this a number of concrete and theoretical ways, examining patterns in snowdepth, snowfall, and temperature as predictors in ski resort attendance (Hamilton, Brown, & Keim, 2007); or suggesting a Climate Index for Tourism, which considers temperature, sunshine, and prioritizes the absence of wind and rain (de Freitas et al., 2008).

We operationalize exposure in a measure of overall water balance with a broad, multisector indicator.[4] The Standardized Precipitation and Evapotranspiration Index (SPEI) builds on the Standardized Precipitation Index (SPI) by incorporating temperature values

(Vicente-Serrano, Begueria, & Lopez-Moreno, 2010). The simple calculation is based on the Thornthwaite (1948) model for potential evapotranspiration (PET) in addition to an overall climatic water balance. It uses the difference between the baseline conditions for PET and precipitation to determine how far a particular moving time window deviates from 'normal.' From this value, a wet or dry period can be parsed out from historical data.

Though it is an increasingly common metric in the hydrological and climatological literature, and it has been used as a drought monitoring indicator (Svoboda & Fuchs, 2016), the SPEI has not been used in studies of park tourism to-date. Because it uses a moving time window, SPEI has the advantage of monitoring various trend types and lengths important for different activities (Svoboda & Fuchs, 2016). At the 12-month scale, it captures the severity of dryness or wetness under a water balance model of a particular year's departure from a normal year.

The yearly window covers long-term social dynamics that factor into drought perception (such as increased water demand and reduced availability) but also to retain a demonstrable relationship with other climatological measures (Beguería, Vicente-Serrano, Reig, & Latorre, 2014).[5] As a long-term indicator, it can be used as a measure of the baseline components that recreators consider important for planning trips in advance. When tailored, the tourism-specific vulnerability equation becomes:

$$\text{Change in visitation} = \frac{\text{recreation (dependence on water)} \times \text{SPEI (hydrological cycle)}}{\text{adaptive capacity (the diversity of activities)}}$$

$$(2)$$

Linking climate conditions with park visitation allows for the use of past years as observational points. Our working hypothesis was that changes in SPEI would have a measurable effect on visitation. The study answers the call set forth in Richardson and Loomis (2004) to couple physical changes with social data that reflect the timing and quantity of recreation using a measure of exposure that contains temperature and precipitation in a time- and space-specific way.[6,7] The qualitative portions of this equation account for the non-linear nature of the relationship between exposure to climate extremes and visitation outcomes. As the dependence on water shifts, or adaptive capacity is increased, park visitation is less dependent on physical exposure to climate conditions.

2.2 Economic significance

Changes in temperature and precipitation do not just affect the ability to enjoy an outdoor experience. There are significant economic repercussions to a drop in park visitation. According to one study, tourist visits to Intermountain West parks were estimated to result in $33.90 in spending per person, per day (Kaval & Loomis, 2003). Current estimates of per-visit values are higher, at an average of $114.27 for the Northern Rockies (see Table 1). For an economically stretched NPS that has requested a yearly budgetary increase of $250 million dollars in 2017, this is an important consideration.

The outdoor tourism industry generates a significant portion of the national economy. It brings $646 billion in consumer spending, $39.9 in federal tax revenue, $39.7 billion in state and local revenue, and directly creates 6.1 million direct jobs (Outdoor Industry Association 2016). However, economic performance is not always connected to park visits and

Table 1. National parks and their main activities. (Source: NPS 2016 and web-user observation of each park's suggested activities under 'Things to Do.')

National Park	Activities
Glacier	Hiking, backcountry camping, ranger-led walks and talks, guided tours, Going-to-the-Sun Road, camping, photography, biking, fishing, boating, stock trails, cross-country skiing, river camping by rafting access
Yellowstone	Camping and hiking, boating, bicycling, camping, cross-country skiing and snowshoeing, dayhiking, fishing, guided wildlife viewing tours, picnicking, walks, snowmobile, wildlife viewing
Grand Teton	Backcountry camping, biking, birdwatching, boating/floating, climbing and mountaineering, concessioner activities (guided backpacking, boating, climbing and packing), cross-country skiing and snowshoeing, fishing, ranger programs (hikes), hiking, horseback riding, mountaineering, scenic drives, wildlife viewing
Rocky Mountain	Hiking, scenic drives, wildlife watching, picnicking, ranger-led programs, visitor centers, camping, fishing, horseback riding, wilderness camping

can take on a multi-directional relationship at the community level. For example, in an exceptionally dry year, 1988, Yellowstone's visitation dropped by more than 390,000 people from the previous year (NPS). While tourism spending can fall precipitously in response to a drop in visitation, the local economy sometimes receives an influx of financial resources from other locations. For example, despite the decline in 1988, Schecter (pers. comm. 2015) found that the surrounding counties received additional revenues related to federal wildfire spending.

The effect of climate on tourism is complex and does not fall along a linear pattern. Drier conditions may not be detrimental in all cases. That is to say, drought does not exclude all recreation opportunities. Some higher alpine objectives become more attainable when snowfields are smaller and thunderstorms are diminished (Sturmer pers. comm., 2017). Rafting outfitters in Colorado report high customer satisfaction in exceptionally dry, low-flow seasons like the drought of 2002 (Shrestha pers. comm., 2016). In these years, cautious rafters can safely access new portions of the river. More economically productive sports are not necessarily linked to climate features, either. Cycling, a sport that is not dependent on snowpack, generates more revenue than snow sports at the national level (Outdoor Industry Association, 2013). Overall, tourism revenues may not suffer directly from climate shifts, and the mechanisms behind this reflect the complexity of outdoor tourism as a leisure activity that is subject to personal discretion.

2.3 Non-climatic sensitivity

Sensitivity is represented by 'non-climatic factors that determine the degree to which the recreation/tourism sector can be affected by a drought' (Thomas et al., 2013). In other words, an activity's sensitivity is not only tied to physical features or disturbances; it is a function of expectations and preferences. For example, if a park is not known for boating or water-based tourism in the first place, then it may not be as susceptible to changes in water levels.

Sensitivity measures are determined through stakeholders' and community input to determine how and whether certain impacts are felt. For example, sensitivity can be affected by water appropriation and rights, or it can be affected by other factors such as the diversity of an outfitter or community in engaging in multiple types of activities. Thomas et al. (2013, p. 5) claim that "Water-dependent businesses such as rafting, fishing, and reservoir-based recreation companies should have additional sources of income on which to rely during years when sufficient water supply is not as available." For each

National Park, this may be different, as it would depend on the degree to which popular activities require having an unimpeded water supply. Each park can be distinguished by the major recreational activities that it promotes (See Table 1).

The activities in each park are fairly congruent, with a few differences. For example, Yellowstone offers more emphasis on picnicking and wildlife viewing, while Grand Teton emphasizes backcountry camping and guided concessioner excursion opportunities. Some of the differences in activities stem from what is allowable (e.g. snowmobiling), and what the individual terrain is best suited for (e.g. mountaineering the Grand Teton summit).

2.4 Adaptive capacity in tourism

Adaptive capacity in response to changing conditions can be measured both in infrastructure's ability to adjust, for example by snowmaking, or in recreators' behavioral ability to change, like waiting to ski when conditions are more favorable. Outdoor enthusiasts and land managers alike are optimistic about the ability to tailor activities around localized changes in weather. Some tourists may have high adaptive capacity to select activities that suit the climate conditions and some activities are readily substitutable for others (Gupta et al., 2010). Some tourists may readily go fly fishing on low-flow rivers, while others will be crestfallen to miss out on the opportunity to go whitewater rafting. Activity substitution is one form of adaptation, and it is worth noting that some activities are more adaptive than others. For example, ski areas will have the option of making snow for downhill skiing, while other snow-dependent sports like snowmobiling, cross-country skiing (Scott & McBoyle, 2006), and snowshoeing may not have this option.[8]

Shifts in management practices are more proactive forms of adaptation that can preserve recreators' chosen activities. Some ski resorts are creative in 'harvesting' snow from base area parking lots and moving it to the skiable terrain; this practice is known in the industry as 'snow hauling' or 'snow farming' (Tragethon, 2016). Arizona's Snowbowl resort has become the first ski resort to use recycled waste water to produce all of its artificial snow (Coffey, 2015). Management practices are just one component of adaptation in the sector, though, as recreators are at a luxury to not passively accept deteriorating conditions. If the weather is not favorable for enthusiasts to participate in a particular activity on a certain day, they are free to choose another.[9]

3 Methodology

3.1 Study sites

The U.S. Rocky Mountain region stands out, in particular, for its mountain ranges that stretch out along the Colorado Front Range, to the strikingly jagged peaks of the Tetons that jut up from the high plains of Wyoming, to the numerous white-capped peaks in Montana. Four iconic U.S. National Parks stand out in the region: Glacier, Yellowstone, Grand Teton, and Rocky Mountain. Figure 1 displays their location. There are many reasons why these parks are appealing: including opportunities to hike, backpack, camp, hone photography skills, bicycle, fish, boat, cross-country ski, picnic, snowmobile, watch wildlife, go for scenic drives, mountaineer, rock climb, and raft.

Northern Rockies National Parks

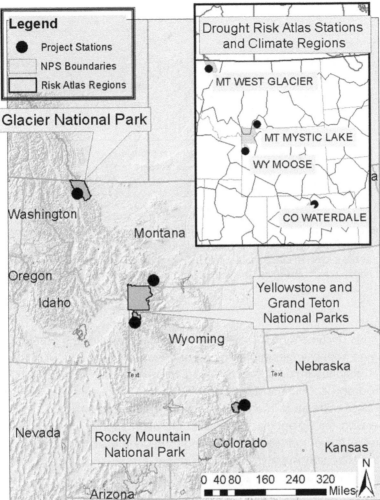

Figure 1. Map of the intermountain region, National Parks, and weather stations used in this study. Data sources: National Parks Service, U.S. Global Historical Climatology Network, and U.S. Census. Weather station locations and Risk Atlas climate regions from the National Drought Mitigation Center.

Residents and visitors to Montana, Colorado, and Wyoming share a similar culture of enthusiasm for outdoor activities, and their economies reflect a larger portion of income from tourism dollars each year. In Montana, outdoor tourism generates $5.8 billion in consumer spending, 64,000 jobs, $1.5 billion in wages and salaries, and $403 million in state and local taxes (Outdoor Industry Association, 2016). In Wyoming, it generates $4.5 billion in consumer spending, 50,400 in direct job creation, and $1.4 billion in wages and salaries (Outdoor Industry Association, 2016). In Colorado, the figures are higher: $13.2 billion in consumer spending, 125,000 jobs created, $4.2 billion in wages and salaries, and $994 million in state and local revenues (Ibid). Climate conditions can have an effect on the economic vitality of outdoor tourism in these states.

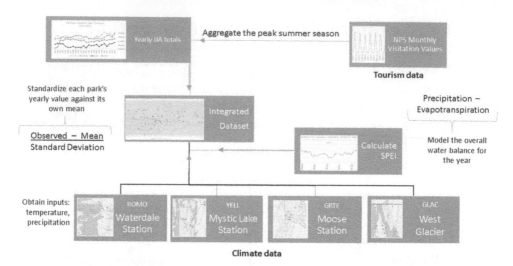

Figure 2. Concept map of data acquisition techniques and processing steps. Beginning with the raw climate data, the indicator length decision was made in advance based on the lag time between onset and impacts for the sector. The aggregate visitation months were selected based on the peak visitation timing and holiday travel schedules. The integrated data-set has observations for each year in each park for 1991–2012.

State reports document how climate variability has already contributed to shifts in the region, ranging from changes in phenology, or the onset and length of seasons; as well as hydrological changes in peak runoff timing from mountain streams in Colorado and Wyoming (Gordon & Ojima, 2015; Riginos et al., 2015). These ecosystem changes can affect tourists who are most interested in spending time outdoors. For example, in Colorado, the season for ski-optimal snowpack has been shortened, montane wildflowers bloom earlier, and river runners reduce rafting trips when peak flows happen earlier in the year, or flows stay consistently lower throughout the season (Jedd, Seidl, & Wilhelmi, 2015; Wilhelmi, Hayes, & Thomas, 2008).

3.2 Data selection and processing

The initial goal was to identify data sources to establish the link between physical and socio-economic indicators of past droughts in U.S. National Parks in the Northern Rockies. This phase involved a series of conversations with drought researchers, and data collection in the field study area. Interviews with park managers, recreation industry representatives, researchers, climate advocacy campaign managers, and recreators in the greater Yellowstone and Grand Teton National Parks region revealed some important themes that were used to refine the research questions and appropriate data sources. To provide quantitative measurements, data sharing was established with the NPS statistics office to obtain visitation numbers for each of the parks. Economic data for the financial impact of visitation was obtained from the most recent NPS Visitor Use Spending report (Thomas & Koontz, 2016).

A dataset pairing visits and the climate indicator allowed for the second phase examining patterns of park visitation under various climate conditions. We use park visitation

Table 2. Concept definitions and measurement. Definitions adapted from IPCC (2014) and measurements from Thomas et al. (2013).

Definitions	Measurement focal points
Exposure: the amount and rate of physical change; or the extent that a system is exposed to significant climate variations (Füssel & Klein, 2006)	Hydrological network, ecosystem health, seasonality, hydrologic cycle, elevation, precipitation, temperatures, streamflow
Sensitivity: the non-climatic features of a system that are not tied to physical disturbances that contribute to the degree to which a system is affected by changes (Füssel & Klein, 2006)	Diversity of economy; dependency on water supply; water rights; past events; human, natural, and economic resources; number of other recreation opportunities (low–high), average household income
Adaptive capacity: the ability to make adjustments to reduce harm or increase resilience; the actions being implemented to address climate conditions as part of a risk management plan (Adger et al., 2007)	Social networks, education and awareness, marketing strategies, public perception, communication plan, seasonal diversification, integrated and alternate opportunities; # acres of public lands for recreation, funding for parks, education level

numbers obtained from the Visitor Use Statistics Program office of the National Park Service in Fort Collins, CO.[10] The period of record available for monthly values begins in 1991 and runs through the end of 2014. The visitor numbers have been adjusted to control for travel that was not for the purposes of recreation. The peak visitation months of June, July, and August are the focus for the study. The monthly values were summed to create a single value for each calendar year. These values were standardized within each park. Intra-park standardization of the visitation values followed the general formula for calculating a z-score.[11] This transformation was done using statistical software designed for social sciences. Standardizing the observed visits in each year allows for inter-park comparison.[12]

Figure 1 displays the regional climate clusters, and where the stations are located in relation to the parks. Stations with long periods of record near entrances or visitor centers were chosen. Figure 2 displays the data processing procedure in a step-wise fashion. These steps were taken iteratively for each park. It was important to carefully consider the time frame length for the climate indication, as this has been linked to the timing and onset of impacts (Bachmair, Svensson, Hannaford, Barker, & Stahl, 2016). Exploratory analysis was done using different SPEI timeframes of 3, 6, and 12 months to see if there were any differences related to the length of a particular dry period. Expertise and evidence suggest that a longer time period is appropriate for the sector as it captures the preplanning that goes into a national park trip and also the lag time required to register a perceptible change in conditions. The data processing steps are highlighted in Figure 2 and Table 3.

The concept map depicts the steps taken to obtain data for tourist and climate trends, process those values, and produce the data-set (see Figure 3). After examining the scatterplots for the sites individually, they were combined in a single plot. From here, a 95% confidence ellipse was applied to identify points outside the ellipse. This visual technique is used to select the majority of observations in a cluster, and can be adjusted for varying confidence levels (Friendly, Monette, & Fox, 2013). The outliers represent the years with particularly wet or dry conditions that corresponded to higher- or lower-than-normal visitation.

Table 3. The methodological steps to select the weather stations, build the data-set, and plot the analytical quadrant.

(1) Using the National Drought Mitigation Risk Atlas tool, select the weather station that falls within the climate layer 'Homogenous Cluster' that is closest to the main park entrance or visitor center. The web interface is located at http://droughtatlas.unl.edu/Data.aspx.
(2) Obtain the 12-month SPEI values for August in each year for the period 1991–2012.
(3) Sum the visitation values for June, July, and August of each year.
(4) Standardize the visitation values as z-scores of departures from their yearly means.
(5) Plot the point observations as 'parkyears' representing their visitation anomaly and standardized SPEI values.
(6) Apply an ellipse to capture 95% of the data points, which are represented with four symbol types.[a]
(7) An optional label can be used to denote the year of each observation.

[a]The r script for the plot:
$p \leftarrow$ ggplot(NPS_SPEI_2, aes(x = SPEI_12mo_JJA_avg, y = JJA_visit_avg)) +
annotate ("rect", xmax = 0, xmin = -3, ymin = 0, ymax = 3, fill = "grey", alpha = 0.45) +
annotate ("rect", xmax = 3, xmin = 0, ymin = 0, ymax = 3, fill = "grey", alpha = 0.5) +
annotate ("rect", xmax = 0, xmin = -3, ymin = -3, ymax = 0, fill = "grey", alpha = 0.5) +
annotate ("rect", xmax = 3, xmin = 0, ymin = -3, ymax = 0, fill = "grey", alpha = 0.45) +
geom_point(aes(x = SPEI_12mo_JJA_avg, y = JJA_visit_avg, shape = ParkID), size = 2) +
geom_text(aes(label = Year), size = 3, hjust = 0, vjust = 0, check_overlap = TRUE) +
stat_ellipse(aes(x = SPEI_12mo_JJA_avg, y = JJA_visit_avg), linetype = "dashed", level = 0.95) +
theme_bw().
The 'comment' sign, #, can be used to turn the labels on and off.

4 Results

4.1 Climate trends

The 12-month SPEI is a measure of the relative wetness or dryness of the previous calendar year. The monthly values, calculated from the observed temperature and precipitation, are plotted as a time series for all four stations (see Figure 4). The August value for each year is the observation point against which the visitation change is plotted.

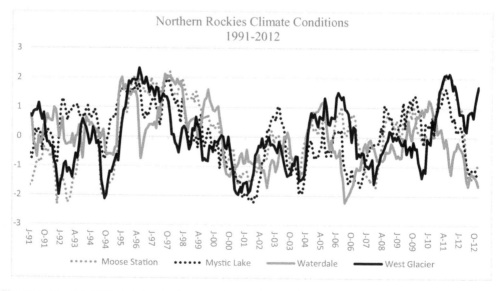

Figure 3. Monthly SPEI plots for the four weather stations in the study. The lower values represent drier years and the higher values represent years with overall wetter conditions. These values are standardized around zero, which is halfway up the y-axis.

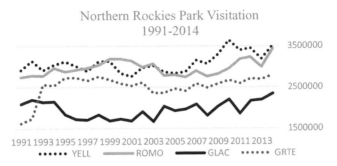

Figure 4. Monthly visitation values for Yellowstone (YELL), Rocky Mountain (ROMO), Glacier (GLAC), and Grand Teton (GRTE).

Dry and wet extremes occurred in each of the four parks during the study period. This is expected, as the SPEI is a calculation of change based on a station's history. Dry and wet periods are a normal part of a location's climatology, and it is not unusual that both are observed during the period of study. During this period, there is a noteworthy convergence in the late 1990s and early 2000s, during which time a multi-year wet period preceded multiple dry years for the study stations.

4.2 Visitation outcomes and economic impacts

In recent decades, yearly visitation has fluctuated in each park, but has generally increased over the years 1991–2014. Intermittent variations in individual park visitation appear throughout the period of record (see Figure 4). These fluctuations show that reductions in one park sometimes coincide with increases in other parks, as with Yellowstone and Rocky Mountain in 2000 and 2001. These patterns are consistent with what Butler (2000) suggests about the nature of the relationship between tourism and the environment: some elements of a vacation may be substitutable and interchangeable when it comes to visitor satisfaction and enjoyment. In other words, visitors may be drawn to visit one park over another based on their own preferences.

The National Park Service (NPS) estimates that in 2015, 3.1 million people visited Grand Teton National Park, an increase of approximately 350,000 from the previous year. It was a year with abnormally high tourism; visitor numbers have not been this high since 1977 and 1978 (National Park Service, 2017). Yellowstone National Park also had a busy year, with 4.1 million people representing an increase of more than 580,000 visitors. Rocky Mountain National Park received 4.2 million tourists, an increase of about 720,000 visitors from the previous year (National Park Service, 2017).

The national economic impacts of park visitation are significant; in 2015, the total 307.2 million visits to all parks and surrounding communities generated $16.9 billion in revenue (Thomas & Koontz, 2016).[13] Out of this total, the Intermountain West Region is a major economic driver, generating the largest percentage of this revenue at $5,994,600. The total visits and the economic data for each of the study parks are listed in Table 4. When each visit is examined in terms of its economic impact, differences between parks are noted. These differences may be attributable to variation in regional and local

Table 4. Park visitation and economic impacts for 2016 (data for columns 1–4 selected from Thomas & Koontz, 2017). The spending per visit was calculated by dividing total visitor spending by the number of visits for each park.

Park	Total recreation visits	Total visitor spending (in thousands, 2016)	Number of jobs supported	Spending per visit
Glacier National Park	2,946,681	$250,815.5	4,337	$85.12
Grand Teton National Park	3,270,075	$597,290.5	9,365	$182.65
Rocky Mountain National Park	4,517,586	$298,746.7	4,575	$66.13
Yellowstone National Park	4,257,177	$524,319.8	8,156	$123.16

economies, as the economic figures represent spending in the surrounding community, hotel stays and meals at restaurants inside and around parks.

4.3 Analysis: the effect of climate extremes

When the climate and visitation data are combined, there are 22 values for each park, with a total of 88 observations. Each park in the data-set is identified with its four letter NPS code. The main cluster of 95% of the observation points is identified inside the dashed line. All parks experienced visitation highs and lows associated with wet and dry conditions.

Figure 5 shows that conditions in 2001 and 2003 in Glacier National Park were dryer than normal by steps between 1 and 2 standardized units below the baseline for the park. At the same time, visitation dropped from the prior year by 48,000 in 2001 and by 241,600 in 2003. However, these years do not fall outside of the ellipse. For Grand Teton, 1991 and

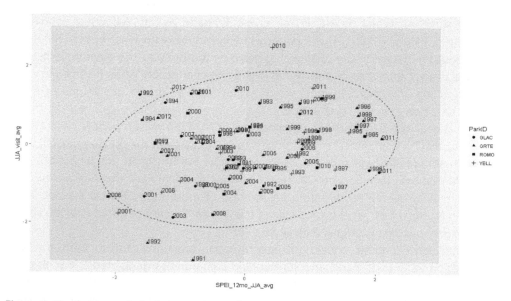

Figure 5. Quadrant graph depicting outlier park-years in wet and dry extremes. Park-code abbreviations are used to indicate which years fell into each category. Yellowstone is abbreviated YELL; Rocky Mountain is ROMO; Grand Teton is GRTE; and Glacier is GLAC and each is denoted by a unique symbol listed in the legend.

1992 stand out as dry years with lower visitation. The largest drop in visits happened in 1991, when there were 494,053 fewer visits than in an average summer. The summer of 1992 was drier and also experienced low visits with 422,255 fewer visits than the mean. Using the per-visit multiplier, the economic value of this decline is significant. These two outliers resulted in an average revenue reduction of $83.7 million per year.

The results show that in response to changes in climate factors, park visitation may not be as hard hit in wet extremes as in dry extremes. The outliers do not fall evenly in the quadrants, with four cases of lower than normal visitation in dry years, out of eight total outliers.[14] Furthermore, the tendency for low visitation outliers to occur in dry years is noteworthy. Of the 16 cases of decreased visitation that are more than one standardized step below normal, 12 of them fall in drier-than-average years. For the years examined, the odds are 75% that a large decline in visitation will happen during a dry year.

The findings of this categorical count fit with what is known about the sector. Tourism takes place during leisure time, so personal comfort plays a large part in whether people chose to spend time being active (Scott, McBoyle, & Schwartzentruber, 2004). Comfort levels are subjective measures based on human perception, and are bounded by the particular environment and context (Zhang, 2016). Many of the climate-related impacts result from drought, or the combination of precipitation deficiency, high temperatures, low humidity, and increased evaporation and transpiration (Wilhite, Sivakumar, & Pulwarty, 2014).

Table 5 displays the results of the analysis for the four quadrants. The final column shows a calculation of the economic impact of the changes in each of the outliers. The four parks generally fared well in wet years, evidenced by the absence of extreme outliers in the lower right quadrant. The borderline outlier is Glacier 2011. Given the focus on trial-hiking, it makes sense that wet conditions would have a negative effect on visits. Images and meteorological studies paired with narratives from park visitors' support this finding as well; for example, wet conditions in 2006 led to flooding at Many Glacier Lodge to the point of road damage and closure (Bernhardt, 2006). In early summer, snow can also impede visitor road access. Even when travelers can access the park where roads have not been damaged or closed, recreation experiences can be less than optimal. For example, in 2011, two backpackers describe being tent-bound during June rainstorms to the point where the experience did not meet their hopes (Lyons, 2011). This and other anecdotes offer a window into the connection between weather conditions and tourism. Even

Table 5. Outlier cases in extreme wet and dry conditions classified by park-year, overall wet or dry status, and whether visitation increased or decreased.

	Park	Outlier Park-Year	12- month SPEI	Relative change in visitation	Change in visits	Estimated change in spending (in 2016 USD)
(1)	Yellowstone	YELL 2001	Dry	−	274,490	−$33,806,188
(2)	Grand Teton	GRTE 1991	Dry	−	494,053	−$90,238,780
(3)	Grand Teton	GRTE 1992	Dry	−	422,255	−$77,124,876
(4)	Rocky Mountain	ROMO 2006	Dry	−	130,664	−$8,640,810
(5)	Yellowstone	YELL 2012	Dry	+	219,519	$27,035,960
(6)	Glacier	GLAC 1992	Dry	+	203,367	$17,310,599
(7)	Yellowstone	YELL 2010	Wet	+	386,536	$47,605,774
(8)	Glacier	GLAC 2011	Wet	−	108,494	−$9,235,009

though it is a marginal outlier, when the aggregate of this effect is taken into consideration, the drop in visits in Glacier 2011 represents 108,494 less than average. When using the per visit spending multiplier, this would have resulted in a drop in revenue of $9,235,009 in 2016 USD.

5 Discussion

Using a climate indicator in combination with visitation numbers allows for a comparison of vulnerability across the four sites in wet and dry years. While the approach has some shortcomings in that it does not account for non-climatic drivers, it begins the process of identifying how substitution patterns emerge between the parks when focused on a specific goal of evaluating climate effects. It also provides a simple way to measure the economic effects of observed changes in visitation that coincided with climate extremes in past years.

The results reveal qualities about the exposure – impact relationship that warrant further investigation. The combination of heat threshold effects on visitation and the fire risk that follows precipitation shortages may explain why visitation drops more during dry conditions. This warrants investigation to uncover the magnitude of drought's effect on tourism. In order to conduct a thorough risk analysis, it will be important to understand what other factors are driving these changes from a geographic and environmental perspective (Butler, 2000), how economic variables affect key outcomes (Stevens et al., 2014), and the degree to which populations are susceptible to harm from these shifts (Hayes et al., 2004).

According to a risk assessment framework for drought (Wilhite et al., 2000), vulnerability is just one component of determining risk. Transferring this lesson to the tourism sector means that in addition to the basic meteorological barriers to recreation, it is important to understand shifts as they relate to changes in activity preference and the dimensions of governance in a particular sector (Miller et al., 2010). Changes in tourist behavior may be driven by an interplay of climatic, social, and political factors that are difficult to parse out. For example, a dry year in Denver, Colorado may result in higher visitation in Rocky Mountain National Park as visitors seek cooler, high alpine environments.

Furthermore, the climate signal may be complicated by the adjacency of public and private lands, seasonality, and management choices. During February 2015, staff in Grand Teton noted that snow was reduced in the nearby ski areas, observing the number of people visiting the Park was likely higher, 'Because of the lack of snow in the valley, commercial ski resorts/facilities are having a lackluster season. This has resulted in more day users entering the park this month' (National Park Service, 2017). In this case, was visitation driven by management decisions surrounding comparatively proactive trail maintenance in the park or insufficient snowmaking in the nearby ski resorts? This is worth exploring in future research. A limitation of this study is that it only considers tourism on public lands, even though resorts and private lands are recreational spaces as well. When viewed as complementary spaces, public and private lands can be used next to one another as destinations during changing seasons and conditions.

5.1 The human dimension of climate vulnerability

The social vulnerability framework calls for a closer consideration of the effects that natural conditions have on humans.[15] It lends to an understanding of how tourism opportunities persist under climate extremes. Keeping this in mind, the next generation of tourism planners could more carefully consider how environmental hazards pose a risk to specific types of recreation. This supports the general sentiment behind the Fifth Annual IPCC report, which emphasizes that it is important to understand the physical science behind the risks of climate change, because this will improve knowledge of how we should prioritize those risks.

Assuming tourists can adjust to changes in climate, though, there are some further considerations concerning the variation between how vulnerable different groups will be. As the IPCC states 'climate change will amplify existing risks and create new risks for natural and human systems. Risks are unevenly distributed and are generally greater for disadvantaged people and communities in countries at all levels of development' (IPCC, 2014). At the more local level, this same logic applies, and just as disadvantaged groups exist within individual countries, some parks may be more vulnerable to climate change than others.[16]

It will also be important to have an updated understanding of changes in recreator preferences, motivations, and how opportunities meet their criteria under climate extremes. To uncover *in situ* preferences, for example, survey instruments could clarify parts of the decision-making process that are not apparent in visitation data alone. One example is a contingent visitation approach that links visitor preferences to conditions under future climate scenarios (Richardson & Loomis, 2004).[17]

Given that alpine national parks are spaces where outdoor recreation is prioritized (see Table 2), tourism may be more dependent on weather and climate in the Rocky Mountains than in other parts of the United States. Extreme climate conditions may adversely affect the volume of park visits. However, this does not mean that park managers have to accept a decline in visitation, particularly under hot or dry extremes. Since a vulnerability analysis points to the sensitivity of systems, in addition to exposure, it allows researchers to identify potential risk based on non-climatic features such as personal preferences, social factors, and built infrastructure (Blaikie et al., 1994). These other factors are worth considering in the context of park management. Future work to understand the vulnerability of tourism to climate extremes would be wise to consider a multiplicity of drivers, including costs of park visits relative to overall prosperity and family incomes. It is quite plausible that some of the variation in visitation outcomes is due to an interplay of economic factors beyond climate conditions considered here.

Furthermore, understanding localized impacts of climate does not tell managers what decisions to make; in fact, a meta-study found that management recommendations are based on ecological reasoning rather than individual, empirical cases, and they are geared toward general audiences and not particular landscapes (Heller & Zavaleta, 2009). If planning is to become more site-specific, it would be beneficial to connect finer-scale visitation data (e.g. at the daily level) with station data (that is available at this resolution). Once these effects are known, it may be possible to tailor management recommendations. Having stable infrastructure in place may position parks to be more resilient during wet years. Paved roads for scenic driving, and wooden boardwalks and elevated paths provide access to geothermal viewing opportunities, which rank as the top two most popular activities in the park (Jorgenson & Nickerson, 2016). To create fulfilling experiences for adventure-seekers, it is crucial to secure basic infrastructure in the parks where visitors use foot pathways that can be damaged by soil erosion following a drought, or washed away during excessive rains. Paved walkways and boardwalks keep tourists safe when hiking near cliffs or thermal pools that can burn them. Furthermore, staying on trails is wise to minimize dangerous interactions with wildlife that may be aggravated by weather conditions.[18]

In addition to ensuring access through roads and pathways, there may be a need for managers in other parks to plan for more safe adventure choices (Manning & Anderson, 2012). For example, systems like *via ferrata* climbing routes have backup gear that prevents scramblers from falling long distances. These systems keep extreme hikers clipped in with safety harnesses, ensuring safety in wetter conditions. Some of the same measures that keep tourists safe during extreme climate conditions may be useful in achieving more general goals in parks, too. In this sense, adaptation need not be viewed as an excessive burden.

6 Conclusion

To date, little work has been done to estimate the economic impacts of climate extremes on national park visitation. While Fisichelli et al. (2015) considered temperature effects on visitation, this study expanded that to examine whether temperature and precipitation extremes affected visitation in four national parks in the northern Rocky Mountains. The study was conducted to improve our understanding of how alpine park tourism and comprehensive climate conditions may be connected. Vulnerability research is actor- and action-oriented, with the goal of preparing for change, so we have also offered examples of adaptive measures. Others emphasize the importance of using a participatory and inclusive approach to develop indices in a way that is relevant for a particular region or set of stakeholders (Thomas et al., 2013). However, prior research had not always clearly spelled out how this could be done. This study offers a beginning framework for examining climate features as a preliminary step in the process.

The outlier analysis shows that park visitation responds differently to climate conditions. Six out of the 8 cases occurred during drier than normal or drought conditions, and 4 out of these 6 resulted in a decrease in visitation. These drops in visitation are associated with subsequent revenue loss for parks and surrounding areas. The scattered nature of the data points suggests that recreation activities are not all affected in the same way by climate conditions. However, when it comes to large-scale changes, drought may have more of a connection to visitation declines. Keeping this in mind, management decisions around water conservation and fire prevention may mitigate some of the visitation decline observed in dry years. Outcomes of extreme conditions may vary depending on the features of a particular park, and the ability of managers to cope with risks related to reduced water availability and increased fire danger. These park-level variations provide a window for tourist operators to learn from one another, as they make adjustments to park infrastructure that reduce risk to tourists.

The sheer variety in the sector means that the relationship between climate factors and tourist opportunities is not linear and depend on the region, access/amenities, and the particular activity. Consideration of personal preferences, individual attitudes and perceptions is another important factor to consider in the recreation sector. As a subset of climate research, work on drought has been critically attuned to the social dimension. Drought can be defined as insufficient water to meet demand (Glantz & Katz, 1977; McNutt et al., 2013). Consideration of additional non-climatic components will build a more robust park tourist model to parse these effects. Future work should consider inter-regional and inter-seasonal comparability in park-based recreation – e.g. in other parks, during the winter, as well as the fall and spring shoulder seasons.

Notes

1. Building from previous studies, an inclusive approach should develop a sector-specific research framework relevant for a particular region or set of stakeholders.
2. It is worth mentioning de Freitas (2003) acknowledged that selecting the basic outcome variable is only the first step, and additional research is warranted to develop a sector-specific design that would integrate the values with relevant climate aspects.
3. Involving stakeholders in the process improves the accuracy of research products (Pentland, 2014), and can also lead to improved compliance outcomes because participants develop their own monitoring guidelines. The recreation industry often participates in drafting drought preparedness plans in Colorado where two of the major stakeholders are Vail Resorts and the Colorado River Outfitters Association (Colorado Water Conservation Board, 2013).
4. Data constraints are more than a nuisance as measures of social phenomena rarely match the scale or units of the climatological data. When studying these effects over time, it becomes difficult to place trends in social and physical processes into comparable terms. For the sake of simplicity, a single indicator is needed that includes changes in both temperature and precipitation.
5. Following a global level analysis of time frames, increasing the length beyond the12-month SPEI was not shown to have improved accuracy (Beguería et al., 2014).
6. For reasons largely related to data constraints, comprehensive work on full climate features has not yet been done for Northern Rockies National Parks.
7. Since the combined effect of temperature and precipitation play a known role in tourist comfort, the study connects temperature and precipitation to Richardson and Loomis' call to examine physical factors more closely.
8. One possible way private recreation industry leaders can do this may be to alter a marketing strategy or diversify the activities a company is providing. Recreation on public lands may have similar adaptive capacity.
9. Study limitations include the inability to design a quantitative metric for the amount of adaptive capacity in a system.
10. Other studies have used the information gathered from working with resort personnel on daily attendance, such as skier and snowboarder visits (Hamilton et al. 2007). This has also been modeled for future IPCC scenarios. This 'contingent visitation analysis' considers the effects of various hypothetical climate scenarios (like changes in temperature, precipitation, and snow depth) on the likelihood of people visiting national parks (Richardson & Loomis, 2004).
11. $Z_X = \frac{X_i - \bar{X}}{S_X}$. The X signifies the visitation variable with i as the observed value, mu or X-bar as the mean, and S as the standard deviation.
12. Because Yellowstone and Rocky Mountain receive higher numbers of visits overall, it would not make sense to compare their changes in visitation to Glacier and Grand Teton.
13. These figures are for gateway regions, which Thomas and Koontz (2016) define as the 60-mile area surrounding parks.
14. The results were contextualized with comments entered by park data technicians. Some of these comments are presented in the results and discussion as evidence to provide context for the observed trends.
15. This fits with Burton, Kates, and White (1993), who argue that choice, human agency, and a consideration of individual decisions should determine the most effective responses to climate extremes.
16. This is especially important in the context of the 2015 Paris convention and follow-up 2016 Marrakech meetings, when countries formulated climate action plans aimed at meeting reductions targets.

17. Richardson and Loomis (2004) consulted with a team of experts to understand the effects that projected changes (according to global climate model scenarios) would have on a national park. Changes included the number of days that trails remained snow-free for hiking, the number of days a high-elevation road would remain open, wildlife abundance, and vegetation composition. From these, a survey asked park visitors how likely they would be to visit, or whether they would shorten or lengthen their stay. Generally speaking, they find the impacts from these climate scenarios are less likely to prevent visitors from coming to a park, since they are planned well in advance. However, the length of stay was adjusted. In particular, extreme heat has a particularly noteworthy effect (Richardson & Loomis, 2004). This fits with what we would expect, given that recreation is voluntary and done for enjoyment (de Freitas, 2003).
18. Park recreators might consider waterway safety during peak runoff in high flow years, and also in low flow years when new rock features can emerge.

Acknowledgements

The authors would like to acknowledge the contributions of Drs Olga Wilhelmi and Andy Wood, project scientists at the Research Applications Laboratory at the National Center for Atmospheric Research. Their input was fundamental in the early conceptual development and measurement of the socioeconomic variables and climate inputs, respectively. Dr Wonho Nam, assistant professor at Hankyong National University, Kyonggi-do, Korea, provided input for the suite of climate monitoring indices available; his guidance on the SPEI was especially helpful.

Disclosure statement

No potential conflict of interest was reported by the authors.

References

Adger, W. N., & Kelly, P. M. (1999). Social vulnerability to climate change and the architecture of entitlements. *Mitigation and Adaptation Strategies for Global Change 4*(3), 253–266.

Adger, W. N., Agrawala, S., Mirza, M. M. Q., Conde, C., O'Brien, K. L., Pulhin, J., Pulwarty, R., Smit, B., & Takahashi, K. (2007). Assessment of adaptation practices, options, constraints and capacity. In Climate Change 2007: Impacts, Adaptation and Vulnerability (Ed.), Contribution of Working Group II to the Fourth Assessment Report of the Intergovernmental Panel on Climate Change (pp. 719–743). Cambridge: Cambridge University Press.

Agnew, M. D., & Viner, D. (2001). Potential impacts of climate change on international tourism. *Tourism and Hospitality Research, 3*(1), 37–60.

Bachmair, S., Svensson, C., Hannaford, J., Barker, L. J., & Stahl, K. (2016). A quantitative analysis to objectively appraise drought indicators and model drought impacts. *Hydrology and Earth System Sciences, 20*, 2589–2609.

Begueria, S., Vicente-Serrano, S., Reig, F., & Latorre, B. (2014). Standardized precipitation evapotranspiration index (SPEI) revisited: parameter fitting, evapotranspiration models, tools, datasets and drought monitoring. *International Journal of Climatology, 34*(10), 3001–3023.FV

Bernhardt, D. (2006). Glacier *national park flooding November* 2006: Attachment 08–23. Great Falls, MT: NOAA/NWS. Retrieved from https://www.weather.gov/media/wrh/online_publications/talite/talite0823.pdf

Blaikie, P., Cannon, T., Davis, I., & Wisner, B. (1994). *At risk*. London and New York, NY: Routledge.

Burton, I., Kates, R. W., & White, G .F. (1993). *The environment as hazard* (2nd ed.). New York, NY /London: The Guilford Press.

Butler, R. (2000). Tourism and the environment: A geographical perspective. *Tourism Geographies, 2* (3), 337–358.

Chape, S., Spalding, M., & Jenkins, M. (2008). *The world's protected areas: Status, values and prospects in the 21st century*. La Mancha: Univ de Castilla.

Coffey, H. (2015, April). Skiing on sewage: Resorts use treated wastewater for snow cannons. *UK Telegraph*, p. 17. Retrieved from http://www.telegraph.co.uk/travel/ski/news/Skiing-on-sewage-resorts-use-treated-wastewater-for-snow-cannons/

Colorado Water Conservation Board (2013). *2013 state drought mitigation & response plan*. Colorado Department of Natural Resources. Retrieved from http://cwcb.state.co.us/water-management/drought/Pages/StateDroughtPlanning.aspx

de Freitas, C. R. (2003). Tourism climatology: Evaluating environmental information for decision making and business planning in the recreation and tourism sector. *International Journal of Biometeorology, 48*, 45–54.

de Freitas, C. R., Scott, D., & McBoyle, G. (2008). A second generation climate index for tourism (CIT): Specification and verification. *International Journal of Biometeorology, 52*(5), 399–407.

Dale, D., & Weaver, T. (1974). Trampling effects on vegetation of the Trail Corridors of North Rocky Mountain Forests. *Journal of Applied Ecology, 11*(2), 767–772.

Delgado, E. (2016, November 17). Personal Communication. Boise, ID: National Interagency Fire Center.

Dow, K. (2010). News coverage of drought impacts and vulnerability in the US Carolinas, 1998–2007. *Natural Hazards, 54*(2), 497–518.

Eagles, P. F. J. (2014). Research priorities in park tourism. *Journal of Sustainable Tourism, 22*(4), 528–549.

Eakin, H., & Luers, A. L. (2006). Assessing the vulnerability of social-ecological systems. *Annual Review of Environmental Resources, 31*, 365–394.

Fisichelli, N. A., Schuurman, G. W., Monahan, W. B., & Ziesler, P. S. (2015). Protected area tourism in a changing climate: Will visitation at US National Parks warm up or overheat? *PLoS ONE, 10*(6), 1–13.

Franke, M. A. (2000). *Yellowstone in the Afterglow: Lessons from the fires: YCR- NR-2000-03.* WY: National Park Service, Mammoth Hot Springs. Retrieved from https://www.nps.gov/yell/planyour visit/upload/full-2.pdf

Frazier, K. (1979). *The violent face of nature: Severe phenomena and natural disasters.* New York, NY: Morrow.

Friendly, M., Monette, G., & Fox, J. (2013). Elliptical Insights: Understanding Statistical Methods Through Elliptical Geometry. *Statistical Science, 28*(1), 1–39.

Füssel, H.-M., & Klein, R. J. T. (2006). Climate change vulnerability assessments: An evolution of conceptual thinking. *Climatic Change, 75*(3), 301–329.

Glantz, M. H., & Katz, R. W. (1977). When is a drought a drought? *Nature, 267*(5608), 192–193.

Gochis, D., Schumacher, R., Friedrich, K., Doesken, N., Kelsch, M., Sun, J., … Brown, B. (2015). The great Colorado flood of September 2013. *Bulletin of the American Meteorological Society, 96*(9), 1461–1487.

Goetz, S. J., Jantz, P., & Jantz, C. A. (2009). Connectivity of core habitat in the Northeastern United States: Parks and protected areas in a landscape context. *Remote Sensing of Environment, 113*(7), 1421–1429.

González Tánago, I., Urquijo, J., Blauhut, V., Villarroya, F., & De Stefano, L. (2016). Learning from experience: A systematic review of assessments of vulnerability to drought. *Natural Hazards, 80*(2), 951–973.

Gordon, E., & Ojima, D. (2015). *Colorado climate change vulnerability study: A Report submitted to the Colorado energy office.* Boulder and Fort Collins, CO. Western Water Assessment, University of Colorado - Boulder, and Colorado State University. Retrieved from http://wwa.colorado.edu/climate/co2015vulnerability/co_vulnerability_report_2015_final.pdf

Gupta, J., Termeer, C., Klostermann, J., Meijerink, S., van den Brink, M., Jong, P., … Bergsma, E. (2010). The adaptive capacity wheel: A method to assess the inherent characteristics of institutions to enable the adaptive capacity of society. *Environmental Science & Policy, 13*(6), 459–471. doi:10.1016/j.envsci.2010.05.006

Hall, C. M., & Page, S. J. (1999). *The geography of tourism and recreation: Environment, place, and space.* New York, NY: Routledge.

Hamilton, L., Brown, C., & Keim, B. D. (2007). Ski areas, weather and climate: Time series models for New England case studies. *International Journal of Climatology, 27*, 2113–2124.

Hayes, M. J., Wilhelmi, O. V., & Knutson, C. L. (2004). Reducing drought risk: Bridging theory and practice. *Natural Hazards Review, 5*(2), 106–113.

Heller, N. E., & Zavaleta, E .S. (2009). Biodiversity management in the face of climate change: A review of 22 years of recommendations. *Biological Conservation, 142*(1), 14–32.

Intergovernmental Panel on Climate Change (IPCC). (2014). *Climate change 2014 Synthesis report summary for policymakers.* Geneva, Switzerland. Retrieved from http://www.ipcc.ch/pdf/assessment-report/ar5/syr/AR5_SYR_FINAL_SPM.pdf

Jedd, T., Seidl, A., & Wilhelmi, O. (2015). *"Climate vulnerability* in Colorado's outdoor tourism and recreation sector." In E. Gordon, & D. Ojima, (Eds). *Colorado Climate Vulnerability.* F ort Collins, CO: Western Water Assessment and Colorado State University.

Jiang, N., Martin, S., Morton, J., & Murphy, S. (2015). *"The bathtub ring. Shrinking lake mead: Impacts on water supply, hydropower, recreation and the environment." Faculty advisor: Naomi Tague. Colorado governance initiative.* Getches-Wilkinson Center for Natural Resources, Energy, and the Environment. Boulder, CO: University of Colorado Law School.

Jones, B., & Scott, D. (2006). Climate change, seasonality and visitation to Canada's National Parks. *Journal of Park and Recreation Administration, 24*(2), 42–62.

Jorgenson, J. D., & Nickerson, N. P. (2016). *Building constituency at Yellowstone National Park: Predicting visitor support now and into the future. Paper 338.* Missoula, MT: University of Montana. Retrieved from http://scholarworks.umt.edu/cgi/viewcontent.cgi?article=1336&context=itrr_pubs

Kaval, P., & Loomis, J. (2003). *Updated Outdoor Recreation Use Values with Emphasis on National Park Recreation: Report prepared for Dr. Bruce Peacock, National Park Service, Fort Collins, CO under*

Cooperative Agreement CA 1200-99-009, Project number IMDE-02-0070. Fort Collins, CO. Retrieved from https://www.researchgate.net/profile/John_Loomis3/publication/237268913_Updated_Out door_Recreation_Use_Values_with_Emphasis_on_National_Park_Recreation/links/552d16da0c f21acb09212d01.pdf

Klein, R. (2011). *Outdoor recreation sector in Colorado: Chapter 7, Colorado Climate Preparedness Project Final Report*. Boulder, CO. Retrieved from www.coloadaptationprofile.org

Knowles, T., & Colwell, R. (2012). *Revisiting Leopold: Resource stewardship in the National Parks: A report of the National Park System Advisory Board Science Committee*. Washington, DC: National Park Service. Retrieved from https://www.nps.gov/calltoaction/PDF/LeopoldReport_2012.pdf

Lambert, A. M., Miller-Rushing, A. J., & Inouye, D. W. (2010). Changes in snowmelt date and summer precipitation affect the flowering phenology of Erythronium grandiflorum (glacier lily; Liliaceae). *American Journal of Botany, 97*(9), 1431–1437.

Loomis, J. (2008). *Economic Development Report – The economic contribution of instream flows in Colorado: How angling and rafting use increase with instream flows (EDR 08-02, Final Report)*. Fort Collins, CO. Retrieved from http://www.fcgov.com/nispreview/pdf/loomis_report.pdf

Loomis, J. B., & Keske, C. M. (2009). Mountain substitutability and peak load pricing of high alpine peaks as a management tool to reduce environmental damage: A contingent valuation study. *Journal of Environmental Management, 90*(5), 1751–1760.

Lyons, C. (2011). National Parks: Glacier. *Backpacker, June*. Retrieved from http://www.backpacker.com/trips/montana/glacier-national-park/national-parks-glacier/#bp=0/img1

Manning, R. E., & Anderson, L. E. (2012). *Managing outdoor recreation: Case studies in the National Parks*. Cambridge, MA: CABI Press.

Martín, M.B.G. (2005). Weather, climate and tourism a geographical perspective. *Annals of Tourism Research, 32*(3), 571–591.

McDougall, K. L., Khuroo, A. A., Loope, L. L., Parks, C. G., Pauchard, A., Reshi, Z. A., ... Kueffer, C. (2011). Plant invasions in mountains: Global lessons for better management. *Mountain Research and Development, 31*(4), 380–387.

McFarlane, B., & Witson, D. (2008). Perceptions of ecological risk associated with mountain pine beetle (Dendroctonus ponderosae) infestations in Banff and Kootenay National Parks of Canada. *Risk Analysis, 28*(1), 203–212. doi:10.1111/j.1539-6924.2008.01013.x

McNutt, C. A., Hayes, M. J., Darby, L. S., Verdin, J .P., & Pulwarty, R. S. (2013). Developing early warning and drought risk reduction strategies. In L. Botterill, & G. Cockfield (Eds.). *Drought, risk management, and policy: Decision making under uncertainty* (pp. 151–170). Boca Raton, FL: Taylor & Francis.

Miller, F., Osbahr, H., Boyd, E., Thomalla, F., Bharawani, S., Ziervogel, G., ... Nelson, D. (2010). Resilience and vulnerability: Complementary or conflicting concepts? *Ecology and Society, 15*(3), 1–25. Retrieved from http://collections.unu.edu/view/UNU:2112#.VgrCH9pPdPg.mendeley

Monz, C. A., Cole, D. N., Leung, Y.-F., & Marion, J. L. (2010). Sustaining visitor use in protected areas: Future opportunities in recreation ecology research based on the USA experience. *Environmental Management, 45*(3), 551–562.

Morris, D., & Walls, M. (2009). Climate Change and Outdoor Recreation Resources. Washington, DC: Resources for the Future.

National Park Service. (2017). Visitor use statistics: Monthly visitation comments by Park. Retrieved from https://irma.nps.gov/Stats/Reports/Park

Nepal, S. K. (2002). Mountain ecotourism and sustainable development. *Mountain Research and Development, 22*(2), 104–109.

Ostrom, E. (2009). A general framework for analyzing sustainability of Social-ecological systems. *Science, 325*(5939), 419 LP–422. Retrieved from http://science.sciencemag.org/content/325/5939/419.abstract

Outdoor Industry Associatio. (2013). Physical measures of vulnerability The Outdoor Recreation Economy. Boulder, CO. Retrieved from https://outdoorindustry.org/images/researchfiles/OIA_OutdoorRecEconomyReport2012.pdf

Outdoor Industry Association. (2016). Research tools: Outdoor recreation economy. Retrieved from https://outdoorindustry.org/research-tools/outdoor-recreation-economy

Pearson, S. M., Turner, M. G., Wallace, L. L., & Romme, W. H. (1995). Winter Habitat use by large ungulates following fire in Northern Yellowstone National Park. *Ecological Applications, 5*(3), 744–755.

Pederson, G. T., Gray, S. T., Woodhouse, C. A., Betancourt, J .L., Fagre, D. B., Littell, J. S., … Graumlich, L. J. (2011). The unusual nature of recent snowpack declines in the North American Cordillera. *Science, 333,* 332.

Pentland, A. (2014). *Social physics how good ideas spread: the lessons from a new science.* New York, NY: The Penguin Press.

Pielke, R. A., Sr, Doesken, N., Bliss, O., Green, T., Chaffin, C., Salas, J. D., … Wolter, K. (2005). Drought 2002 in Colorado: An unprecedented drought or a routine drought? *Pure & Applied Geophysics, 162*(8–9), 1455–1479.

Proctor, J. (2017). Personal communication. Washington, DC: Public Risk Management, Office of Risk Management. National Park Service. 20 April.

Quammen, D. (2016). It's our land. Let's keep it that way. *New York Times.* Retrieved from http://www.nytimes.com/2016/12/10/opinion/its-our-land-lets-keep-it-that-way

Richardson, R. B., & Loomis, J. B. (2004). Analysis – Adaptive recreation planning and climate change: A contingent visitation approach. *Ecological Economics, 50,* 83–99.

Richardson, R. B., & Loomis, J. B. (2005). Climate change and recreation benefits in an Alpine National Park. *Journal of Leisure Research, 37*(3), 307–320.

Riginos, C., Newcomb, M., Wachob, D., Schecter, J., & Krasnow, K. (2015). *The coming climate: Ecological and economic impacts of climate change on Teton County.* The Charture Institute. Retrieved from http://charture.org/wp-content/uploads/2016/04/The-coming-climate-sept-2015.pdf

Scott, D., & McBoyle, G. (2006). Climate change adaptation in the ski industry. *Mitigation and Adaptation Strategies for Global Change, 12*(8), 1411–1431.

Scott, D., Jones, B., & Konopek, J. (2007). Implications of climate and environmental change for nature-based tourism in the Canadian Rocky Mountains: A case study of Waterton Lakes National Park. *Tourism Management, 28*(2), 570–579.

Scott, D., McBoyle, G., & Schwartzentruber, M. (2004). Climate change and the distribution of climatic resources for tourism in North America. *Climate Research, 27,* 105–117.

Shea, D. (2016). NCAR/UCAR climate data guide: Statistical and diagnostic methods overview. Retrieved from https://climatedataguide.ucar.edu/climate-data-tools-and-analysis

Shrestha, P. (2016, August 15). Personal communication. Lincoln, NE: Sustainability Director, University of Nebraska.

Shrestha, P., & Schoengold, K. (2008). Potential economic impact of drought on rafting activity. Institute of Agriculture and Natural Resources, Agricultural Economics Department. *Cornhusker Economics, Paper 362.*

Stevens, T. H., More, T. A., & Markowski-Lindsay, M. (2014). Declining National Park Visitation: An economic analysis. *Journal of Leisure Research, 46*(2), 153–164.

Sturmer, K. (2017). Personal communication: Rocky Mountain backcountry park ranger (Jedd).

Svoboda, M., & Fuchs, B. (2016). Handbook of Drought Indicators and Indices (WMO-No. 1173). Geneva, Switzerland: World Meteorological Organization and Global Water Partnership.

Thomas, C. C., & Koontz, L. (2016). *2015 National Park visitor spending effects: Economic contributions to local communities, states, and the nation: Natural Resource Report (NPS/NRSS/EQD/NRR-2016/1200).* Fort Collins, CO: National Park Service.

Thomas, C. C., & Koontz, L. (2017). *2016 National Park visitor spending effects: Natural resource report (NPS/NRSS/EQD/NRR—2017/1421).* Fort Collins, CO: National Park Service. Retrieved from https://www.nps.gov/nature/customcf/NPS_Data_Visualization/docs/2016_VSE.pdf

Thomas, D. S. K., Wilhelmi, O., Finnessey, T. N., & Deheza, V. (2013). A comprehensive framework for tourism and recreation drought vulnerability reduction. *Environmental Research Letters, 8,* 1–8.

Thornwaite, C. (1948). An Approach Toward a Rational Classification of Climate. *Geographical Review,* 38(1), 55–94.

Tragethon, D. (2016). Personal communication: Director of Marketing, Mt. Hood Meadows Ski Resort (Smith and Jedd, June 24).

Turner, M. G., Hargrove, W. W., Gardner, R .H., & Romme, W. H. (1994). Effects of fire on landscape heterogeneity in Yellowstone National Park, Wyoming. *Journal of Vegetation Science, 5*(5), 731–742.

U.S. Fish and Wildlife Service. (2015). *U.S. Fish and Wildlife Service: The Cooperative Ecosystem Studies Units (CESU) network*. Washington, DC. Retrieved from https://www.fws.gov/science/pdf/FWS-CESU-FAQs.pdf

Vicente-Serrano, S., Begueria, S., & Lopez-Moreno, J. (2010). A Multiscalar Drought Index Sensitive to Global Warming: The Standardized Precipitation Evapotransipiration Index. *Journal of Climate, 23*, 1696–1718.

Westerling, A .L., Turner, M. G., Smithwick, E. A. H., Romme, W. H., & Ryan, G. M. (2011). Continued warming could transform Greater Yellowstone fire regimes by mid-21st century. *Proceedings of the National Academy of Sciences, 108*, 13165–13170.

Wilhelmi, O., Hayes, M. J., & Thomas, D. S. K. (2008). Managing drought in mountain resort communities: Colorado's experiences. *Disaster Prevention and Management, 17*(5), 672–680.

Wilhite, D. A., Hayes, M J., Knutson, C., & Smith, K. H. (2000). Planning for drought: Moving from crisis to risk management. *JAWRA Journal of the American Water Resources Association, 36*(4), 697–710.

Wilhite, D. A., Sivakumar, M. V. K., & Pulwarty, R. (2014). Managing drought risk in a changing climate: The role of national drought policy. *Weather and Climate Extremes, 3*, 4–13.

Willard, B. E. (1970). Effects of human activities on alpine tundra ecosystems in Rocky Mountain National Park, Colorado. *Biological Conservation, 2*(4), 257–265.

Williams, S. (1998). *Tourism geography*. London and New York, NY: Routledge.

Zhang, T. H. (2016). Weather effects on social movements: Evidence from Washington, D.C., and New York City, 1960–95. *Weather, Climate, and Society, 8*(3), 299–311.

Climate and visitation to Utah's 'Mighty 5' national parks

Jordan W. Smith (iD), Emily Wilkins (iD), Riana Gayle and Chase C. Lamborn

ABSTRACT

The relationship between climate and visitation to managed natural areas has been analyzed at a variety of different spatial scales. We expand upon our existing knowledge on this topic by: (1) determining how a wide range of climate variables affect visitation across a regional tourism system; and (2) identifying which variables affect visitation system-wide and which variables only affect visitation at specific parks. Our analysis focuses on five national parks located in southern Utah (USA) commonly referred to as 'the Mighty 5'. We found monthly average daily maximum temperatures were the best predictor of system-wide visitation, suggesting average daily maximum temperatures play a more direct role in tourists' travel decisions relative to other climate variables, including other derivations of temperature. We also found declines in monthly park visitation for three parks (Arches, Canyonlands, and Capitol Reef) once average daily maximum temperatures exceed 25 C. For Bryce Canyon and Zion however, monthly visitation continued to increase well above this threshold. The geophysical characteristics of these parks appear to mediate the relationship between average daily maximum temperature and visitation. The commonly found 'inverted U-shape' relationship between temperature and visitation should not be seen as a universal maxim. We also found precipitation to be a poor predictor of system-wide visitation, but a significant factor shaping the travel decisions of visitors to Bryce Canyon, the only park to offer snow-based outdoor recreation opportunities. Future research should not disregard the possibility of precipitation being a significant factor shaping visitors' travel decisions. By conducting our analyses at two distinct scales, we have found there is a difference between the individual climate variables that are regionally-significant drivers of visitation and those that are locally-significant drivers of visitation. Scale matters in analyses of the relationship between climate and visitation.

摘要

现有文献已经在不同的空间尺度上分析了气候与自然区域访问量的关系。我们扩展了该主题的现有知识:1)确定广泛的气候变量如何影响区域旅游系统的访问;2)确定哪些变量影响了整个区域旅游系统的访问, 哪些变量只影响特定公园的访问。我们的分析主要集中在美国犹他州南部的五个国家公园, 通常被称为"五强"。我们发现, 月平均最高温度是全区域系统访问的最佳预测指标。这表

(b) Supplemental data for this article can be accessed at (a) https://doi.org/10.1080/14616688.2018.1437767.

明，与其它气候变量(包括其它温度的衍生变量)相比，日平均最高气温对游客的旅行决策起着更直接的作用。我们还发现，一旦月平均最高气温超过25度，有三个国家公园(拱门国家公园、峡谷地国家公园、国会山国家公园)的游客数量会下降。然而，对于布莱斯峡谷和锡安国家公园来说，每月的访问量仍在增加，远超过这个门槛。这些公园的地球物理特征似乎调节日平均最高气温与访问量的关系。常见的"倒U形"曲线不应被视为温度与访问量关系的普遍规律。我们还发现，降水量是整个区域系统访问量的一个糟糕的预测因子，但却是影响布莱斯峡谷游客出行决策的一个重要因素。布莱斯峡谷是唯一一个提供滑雪户外娱乐机会的公园。未来的研究不应忽视降水可能是影响游客出行决策的重要因素。通过在两个不同的尺度上进行分析，我们发现，区域范围影响访问量的重要因素与地方范围影响访问量的重要因素存在差异。尺度在分析气候与访问量关系方面不可忽视。

Introduction

Visitation to managed natural areas is highly dependent upon climate and weather. Many tourists select their destinations based upon expected climatic conditions (Hamilton & Lau, 2006) while many regional tourists and local visitors plan their trips to areas where the near-term weather forecasts project desirable conditions (Patrolia, Thompson, Dalton, & Hoagland, 2017; Rutty & Andrey, 2014). Often, regional and local tourists adjust their trip timing and alter their length of stay or the outdoor recreation activities they participate in, based on the weather (e.g. Becken & Wilson, 2013). The relationship between climate, weather, and visitation to managed natural areas has been analyzed at a variety of different spatial scales ranging from specific national parks (e.g. Richardson & Loomis, 2004; Scott, Jones, & Konopek, 2007), to regional tourism systems (e.g. Coombes, Jones, & Sutherland, 2009; Smith et al., 2016), to national (e.g. Fisichelli, Schuurman, Monahan, & Ziesler, 2015; Liu, 2016) and international (e.g. Barrios & Ibañez, 2015; Lise & Tol, 2002) networks of tourism destinations. These studies most often correlate past visitation rates with a select set of climate variables, among which temperature is used most often (Gössling & Hall, 2006). Here, we expand upon our existing knowledge about how climate and weather affect visitation to managed natural areas by analyzing historical shifts in visitation attributable to a broad set of climate variables across a regional tourism system[1].

Our objectives are twofold: first, to determine how a broad set of climate variables affect visitation to managed natural areas across a tourism system. Analyses of regional, national, and global tourism systems often identify a single climate variable (e.g. average daily mean temperature) that is significantly related to visitation to managed natural areas. Often these studies lack destination-specific data that can be utilized to determine if a wider spectrum of climatic variables (e.g. precipitation, cloud cover, etc.) are also related to visitation to managed natural areas. Exploring how a broad set of climate variables affect visitation across a regional tourism system will improve our understanding of which climate variables are most predictive of visitation. Our second objective is to identify which climate variables affect visitation across an entire tourism system and which climate variables only affect visitation at specific destinations. Many analyses of the relationship between climate and visitation ignore the issue of spatial scale, assuming globally relevant climatic predictors of visitation affect all tourism destinations the same. This may not always be the case; some climate variables that are poor predictors of visitation across an entire system may be highly influential at specific destinations. Similarly, certain climatic conditions that affect visitation system-wide may have only a marginal or negligible effect on visitation at the local level. Scale matters in analyses of climate and visitation to managed natural areas. By identifying which climate variables affect visitation across an entire tourism system and which climate variables only affect visitation at specific destinations, our analyses can illustrate this point.

Related literature

Climate change and tourism

A changing climate has the potential to considerably alter visitation to managed natural areas since outdoor recreationists and tourists are highly sensitive to climate and weather. For example, warming temperatures are likely to decrease the number of days with snow in many locations and thus displace some skiers (e.g. Dawson, Scott, & Havitz, 2013; Rutty et al., 2015; Scott, Dawson, & Jones, 2008). Past research has shown a direct correlation between weather conditions and the closure of New England ski areas (Beaudin & Huang, 2014), illustrating how climate change can alter the economies of tourism destinations. Although there is more research on the impacts of climate and weather on winter outdoor recreation, changes in summer weather have also been shown to influence tourists and thus have an effect on visitation and spending (Denstadli, Jacobsen, & Lohmann, 2011; Falk, 2015).

Previous research has investigated the potential impact of climate change on future national park visitation (e.g. Fisichelli et al., 2015; Liu, 2016; Scott et al., 2007). Scott et al. (2007) modeled future visitation to Waterton Lakes National Park, Canada, by comparing past monthly visitation data to monthly temperature and precipitation. Similarly, Liu (2016) used temperature and precipitation to model visitation to Taiwan's national parks under climate change, finding precipitation was a stronger predictor of visitation. Relatedly, Richardson and Loomis (2004) modeled future visitation to Rocky Mountain National Park (USA), using minimum and maximum temperature, as well as precipitation and snow depth, as predictors of visitation.

Additionally, a previous study investigated the potential effect of climate change on future visitation at all US National Park Service units (Fisichelli et al., 2015). Fisichelli and his colleagues used historical data on visitation and monthly average daily mean temperatures to predict visitation in the future. Results showed that warming temperatures were expected to increase winter visitation for all five national parks in Utah. However, there were differences in projected visitation under warming for the summer months; summer visitation for Zion and Bryce Canyon is expected to stay fairly constant under warming scenarios, while visitation for Arches, Canyonlands, and Capitol Reef is expected to decrease in the summer months (Fisichelli & Ziesler, 2015a, 2015b, 2015c, 2015d, 2015e). These projections only considered monthly average daily mean temperatures as a driving force affecting visitation. We aim to further explore the differing impact of climate on Utah's national parks by investigating how a broad set of variables, including monthly average daily mean temperatures, influence visitation.

Investigating the impact of climate and weather on tourism flows is essential to understanding shifting patterns of visitation under climate change. Tourists can adapt to changing conditions either spatially (changing destinations), temporally (changing trip timing), or behaviorally (changing recreational activities) (Dawson et al., 2013). All of these adaptations impact the management of protected areas as well as the surrounding businesses and communities.

Climate and weather sensitivities for tourism

Four climate and weather variables are commonly studied in tourism research: temperature (minimum, maximum, and/or mean), precipitation, wind, and sunshine (e.g. Hewer, Scott, & Gough, 2015; Scott, Gössling, & de Freitas, 2008; Steiger, Abegg, & Jänicke, 2016). Tourists' perceptions of the importance of these variables depends on the destination's location and the geophysical characteristics of its landscape (Rutty & Scott, 2010; Scott et al., 2008). Surveys of tourists suggest that beach tourists tend to rate sunshine and precipitation as the highest importance (Moreno & Amelung, 2009; Scott et al., 2008), while mountain tourists perceive precipitation to be most influential (Scott et al., 2008; Steiger et al., 2016), while urban tourists are most sensitive to temperature (Scott et al., 2008). Additionally, perceptions of acceptable conditions tend to vary by individual based on their home location, their expectations and experiences of the destination's climate, and their planned recreational activities (Gössling, Abegg, & Steiger, 2016; Rutty & Scott, 2016; Scott et al., 2008). This has led research in dissimilar geographies to reach different conclusions about the impact of climate and weather on tourists. Because surveying visitors is time-intensive and costly, requiring field crews to be on-site at a destination for prolonged periods of time, these types of studies often are conducted on smaller, localized scales.

Research at coarser spatial resolutions often use historical climatic conditions and visitation data to investigate revealed preferences for specific climatic conditions (e.g. Fisichelli et al., 2015; Hewer, Scott, & Fenech, 2016; Jones & Scott, 2006; Loomis & Richardson, 2006; Scott et al., 2007). However, there is little consistency in which climate variables should be included in regional, national, or global modeling efforts. Some studies have focused on only temperature as the sole predictor of visitation, as temperature tends to be correlated with other climate variables and often is the strongest predictor (Bigano, Hamilton, Maddison, & Tol, 2006; Rosselló-Nadal, 2014). However, other studies have shown that additional climate and weather variables are often significant predictors as well (Falk, 2014; Rosselló-Nadal, Riera-Font, & Cárdenas, 2011; Scott & Jones, 2007), and sometimes even more influential than temperature (Yu, Schwartz, & Walsh, 2009). For example, Falk (2014) found that sunshine and temperature both had a positive effect on overnight stays in Austria, while precipitation had a negative effect. Similarly, Yu et al. (2009) found that storms and rain are the most important factors impacting visitors in King Salmon (Alaska, USA) and Orlando (Florida, USA) during the summer. In a different study however, Becken (2013) found that weather did not impact visitation to Westland, New Zealand, across years, although weather did drive tourism seasonality within each year. These differing conclusions suggest that geography and spatial scale may play a substantial role in modeling the impact of climate and weather on tourism.

The effect of climate and weather on tourism has also been studied using climate indices, such as the tourism climatic index (e.g. Amelung & Nicholls, 2014; Amelung & Viner, 2006; Mieczkowski, 1985; Perch-Nielsen, Amelung, & Knutti, 2010; Scott, McBoyle, &

Schwartzentruber, 2004). Tourism indices incorporate multiple climate and weather variables, but often assume the variables specified in the model have the same importance for
every location and tourist type (Scott, Rutty, Amelung, & Tang, 2016). Additionally, the
indices do not account for the local topography of an area. Some destinations may con-
tain more microclimates within one area, which allow for tourists to somewhat alter the
weather they experience (Rutty & Scott, 2014). For example, people in mountainous destinations could travel to higher elevations for cooler weather, or tourists in areas with canyons could seek less sunshine and cooler temperatures by recreating in canyon bottoms.
Some destinations have more microclimates than others, which increases the adaptive
capacity of tourists visiting those destinations.

While a variety of different climate and weather variables have been used in previous
research, the effects of those variables have been inconsistent, appearing to be dependent on the geographic location, dominant activity type at the destination, and spatial
scale of analysis. Studies at the national or international scale tend to only focus on the
importance of temperature, while more site-specific variables are often disregarded (e.g.
Berrittella, Bigano, Roson, & Tol, 2006; Serquet & Rebetez, 2011). However, studies at
smaller spatial scales often find other variables, besides temperature, to be meaningful
predictors of visitation to managed natural areas (e.g. Førland et al., 2013; Köberl,
Prettenthaler, & Bird, 2016; Yu et al., 2009). Because of these inconsistencies, we investigate the impact of numerous climate variables at two spatial scales.

Methods

The Mighty 5

Our analysis focuses on a regional network of national parks located in southern Utah
(Figure 1). The parks – consisting of Arches, Bryce Canyon, Canyonlands, Capitol Reef, and
Zion – are commonly referred 'the Mighty 5'. The moniker is the product of an ad campaign launched by the Utah Office of Tourism in 2013. The campaign sought to market
the parks as a regional tourism destination. Along with the ad campaign, the Utah Office
of Tourism released 3 to 10 day itineraries that would guide visitors from either the Salt
Lake or Las Vegas international airports to two or all five national parks, depending on the
length of the trip (Utah Office of Tourism, N. D.). The success of Mighty 5 was immediately
apparent. Park visitation prior to 2000 had been steadily increasing. In the year the campaign launched, all five of Utah's national parks received a total of 6.3 million visitors
(National Park Service, 2017). After the ad campaign, visitation abruptly rose by 60.3%
over three years, resulting in 10.1 million national park visitors in 2016 (National Park Service, 2017).

To the tourism industry and the state, the increase in visitation and visitor spending
was welcome; however, the millions of additional people visiting the national parks each
year has put a great deal of strain on National Park Service managers. The National Park
Service's mission is twofold: protect the land and serve the people. With millions of additional people coming to Utah's national parks each year, park managers were, and are,
having a difficult time protecting park resources and providing quality national park
experiences.

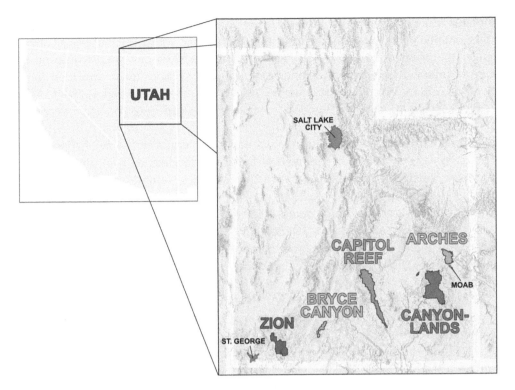

Figure 1. Utah's 'Mighty 5' national parks. Source: Authors.

Characteristics of each national park

Average monthly climatic conditions for each of the five park units are shown in Supplemental Figure 1. Importantly, three of the study parks are located within the Colorado Plateau (Arches, Canyonlands, and Capitol Reef) while the other two parks (Bryce Canyon and Zion) are located in the southernmost extent of the southern Wasatch mountains. The differences in ecoregions affects the types of recreational opportunities that are offered within the parks, with Bryce Canyon and Zion offering more trails in canyons and with vegetative cover. The parks in the Colorado Plateau tend to be more exposed, with few opportunities for escaping daily high summer temperatures that can often exceed 30 °C.

Arches are the most eastern park in the state of Utah, and sits just northeast of Moab. Being part of the Colorado Plateau, the climate can be characterized as 'high desert', which means it is arid, with hot summers and cold winters, and large daily temperature fluctuations that often span a range of 21 °C. The park is characterized by protruding sandstone formations amongst a relatively flat desert floor covered by low-growing vegetation. The difference between the lowest and highest elevations in the park (i.e. vertical relief) is 516 m.

Bryce Canyon is located in southcentral Utah. It has the highest elevation of Utah's national parks (~2778 m), which means it has lower temperatures, more vegetation, and more snow accumulation. Bryce Canyon is the only national park within Utah to offer snow-based recreational activities including cross-country skiing and snowshoeing. The park often receives over 45 mm of precipitation per month in the fall and averages 30 mm of precipitation in the winter months, most of which falls as snow. Much of Bryce Canyon's upper elevations are covered by conifer forests, but as areas of the park descend into lower elevations the vegetation changes into ponderosa pine forest, and then further down it transitions into pinyon and juniper. The vertical relief of the park is 780 m. Across this gradient are hundreds of 'hoodoos', unique steep-sided geological formations that rise from the valley floor.

Canyonlands is located just west of Moab, and is the largest of the five national parks. Like Arches, Canyonlands is in the heart of the Colorado Plateau, giving it many of the same 'high desert' climatic and vegetative characteristics. Summer temperatures in the park often exceed 30 °C. However, just as its name implies, the topography is much more extreme. Total vertical relief of the park is 1030 m. The two main canyons in the park were created by the Green and Colorado rivers, which enter the northern end of the park, converge in the middle, and flow out of the southern end. Many visitors are attracted to Canyonlands due to the kayaking and rafting opportunities offered on both the Green and Colorado River.

Capitol Reef is located in south-central Utah and is characterized by its brightly colored canyons, cliffs, monoliths, and buttes. Maximum daily temperatures often exceed 30 °C in the summer months. The park is centrally located in the Colorado Plateau and its landscape is high arid desert with several slot canyons cut in by the Fremont River. Total vertical relief in the park is 1549 m. The park is filled with several different types of sandstone that comprise hundreds of domes, towers, monuments, bluffs, and spires across the landscape.

Zion is the most southwestern park in Utah and is located at the junction of the Colorado Plateau, the Great Basin, and the Mojave Desert. The landscape is varied, with the lowest point of elevation at 1110 m and the highest at 2660 m. The largest feature in the park is Zion Canyon, which is 15 miles long, and up to half a mile deep. The park's canyons, along with dense vegetative cover in their bottoms, shade and cool many of the most heavily used trails within the park. Shade and cooler temperatures are often a relief, as daily maximum temperatures can often exceed 33 °C in the summer months. Amongst our study parks, Zion offers the largest vertical relief, spanning 1550 m.

All of the study parks have visitor centers which do offer some respite from extreme or unanticipated weather conditions. Given this, the indoor amenities of individual parks are not believed to affect the relationship between specific climate variables and visitation.

Data

Monthly recreation visits

We obtained the total number of monthly recreation visits for the five study parks between January 1979 and December 2014 from the National Park Service's Integrated Resource Management Applications Portal (National Park Service, 2017). A recreation visit is a unique entrance into the park for the purpose of participating in outdoor recreation. The method of counting recreation visits for each park unit is described in Supplemental Table 1.

Monthly visitation by park, averaged between 1979 and 2014, are shown in Figure 2. Because visitation rates are notably different across the five study parks, we converted the

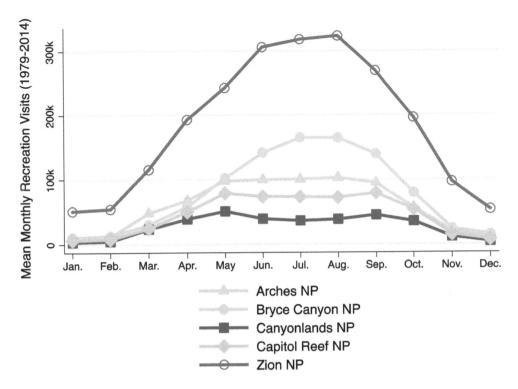

Figure 2. Average monthly visitation for the five study parks (1979–2014). Source: Authors.

raw monthly visitation data to each month's proportion of the yearly visitation total. This monthly proportion served as the dependent variable in the time-series regression models described below. The method of using each month's proportion of the yearly visitation total ensures parks with larger total visitation amounts do not bias estimated coefficients; the method is also consistent with previous research (Fisichelli et al., 2015).

Local climate data

We obtained gridded climate data from the Climatic Research Unit (CRU) time-series version 3.23 (Harris, Jones, Osborn, & Lister, 2014). These data estimate a suite of climate variables at 0.5 decimal degree increments across the globe for every month between January 1901 and December 2014. Given monthly recreation visits for each of the five study parks has only been recorded since January 1979, we only use climate data spanning the same timeframe (January 1979 through December 2014). The suite of climate variables estimated in the CRU time-series data are: monthly average daily mean temperature; monthly average daily minimum temperature; monthly average daily maximum temperature, diurnal temperature range, potential evapotranspiration, precipitation, wet day frequency, frost day frequency, percentage cloud cover; and vapor pressure. Derivations of these data have been used in previous research to examine the climate change exposure of US National Parks (Monahan & Fisichelli, 2014). Additionally, the monthly mean air temperature variable has been correlated with monthly visitation rates to park units across

Table 1. Description of climate variables.

Climate variable	Unit	Description
Monthly average daily mean temp.	Degrees C	The mid-point (i.e. median) between the daily minimum and maximum temperatures.
Monthly average daily minimum temp.	Degrees C	
Monthly average daily maximum temp.	Degrees C	
Diurnal temperature range	Degrees C	The difference between the daily minimum and maximum temperatures
Potential evapotranspiration	mm	The amount of evaporation that would occur if a sufficient water source were available (Bonan, 2002).
Precipitation	mm	
Wet day frequency	Days	Number of days in a month with rain.
Frost day frequency	Day/Month	A frost day is a period of 24 hours in which the minimum temperature falls below 0 degrees C.
Percentage cloud cover		
Vapor pressure	Hecta-Pascals	A measure of liquid's evaporation rate; this increases non-linearly with temperature.

Source: Authors.

the United States (Fisichelli et al., 2015). Details on each of these climate variables are provided in Table 1.

For each study park, we obtained the historical climate data for each grid cell within a 30 km buffer of the park boundary. The use of a 30 km buffer is consistent with previous research (Fisichelli et al., 2015; Monahan & Fisichelli, 2014) and with the NPSs' standards (NPS Natural Resource Inventory and Monitoring Division, 2016). The buffer mitigates against measurement biased introduced from the coarse nature of the climate data and ensures ecological processes beyond a park's boundary, which might also affect visitation, are also captured (Hansen et al., 2011). For each month between January 1979 and December 2014, we averaged the climate variables across each park unit's relevant grid cells to create park-specific average climate measures. Because the parks vary in size and shape, the number of relevant grid cells varied by park (Arches = 2, Bryce Canyon and Zion = 4, Canyonlands = 9, Capitol Reef = 12). The maximum deviations between the averaged values and the observed values for any specific grid cell were: monthly average daily mean temperature = 7.2 °C; monthly average daily minimum temperature = 8.1 °C; monthly average daily maximum temperature = 6.6 °C; diurnal temperature range = 5.6 °C, potential evapotranspiration = 1.6 mm; precipitation = 69.4 mm, wet day frequency = 3.6 days; frost day frequency = 14.9 days; percentage cloud cover = 13.2; and vapor pressure = 3.8 hPa.

US dollar index

International visitation to the United States is driven, to a certain extent, by the value of the US dollar (Anastasopoulos, 1989). Anecdotal evidence suggests national park visitation within the country is also related to the value of the dollar (US Travel Association, 2016). Given this, we also included the US dollar index as an independent variable in our panel time series models. The US dollar index is a measure of the value of the dollar relative to the value of currencies from a set of the country's most significant trading partners. We downloaded US dollar index data from 1 January 1979 to 31 December 2014 from the Federal Reserve (www.federalreserve.gov). We averaged the daily data across the month for inclusion in the regression models described below.

Analysis

Previous climate/visitation research has used average daily mean temperature as a predictor of park visitation (Rosselló-Nadal, 2014). Analysis using average daily mean temperature as the sole predictor of park visitation will not be able to capture shifts in visitation driven by other climate variables (e.g. consistent increases/decreases in precipitation levels or increases in maximum mean air temperature that outpaces increases in mean air temperature). Given this, our analysis included all 10 climate variables within the CRU time-series data-set. We use the raw climate data, as opposed to using a composed climate index (e.g. Mieczkowski, 1985). Due to high correlations between several of the climate variables, each is analyzed in isolation as a distinct predictor of shifts in historical visitation.

We constructed and estimated panel[2] time-series regression models using the monthly climate and visitation data. Macro panel data such as ours often include some form of cross-sectional dependence (CSD), which is simply a lack of independence across units within the sample (Kapetanios, Pesaran, & Yamagata, 2011). More explicitly, CSD results is a correlation structure in the error term across units within the sample attributable to unobservable common factors affecting the relationship between the regressors and the dependent variable. In the case of national park visitation, for example, an unobservable common factor affecting the relationship between climatic conditions and park visitation would be the demographic profile of the domestic visitors. As the US population ages, it is not unreasonable to assume that the relationship between climatic conditions and park visitation may be exaggerated. Lise and Tol (2002) demonstrated how older tourists have different climate preferences relative to younger individuals). Unobservable common factors could include any omitted variable (Banerjee & Carrion-i-Silvestre, 2017). Methods for dealing with CSD have evolved from incorporating time dummies into the regression model to estimating principal components of the residual error terms (Coakley, Fuertes, & Smith, 2002). More recent econometric work by Pesaran (2006) has shown how including cross-section averages in each t of time for both independent and dependent variables can control for unobservable common factors; this is the approach we use in our analysis. Burdisso and Sangiácomo (2016) provide a thorough review of the evolution of panel time series models, with specific attention paid to how to best deal with CSD.

Our model is specified as:

$$y_{it} = \beta_i' \, z_t + \beta_i' x_{it} + \beta_i' d_t + \beta_i' \bar{y}_t + \beta_i' \bar{x}_t + \varepsilon_{it},$$

where y_{it} is each months (t) proportion of its year's annual visitation total for each park (i); z_t is a trend variable, which does not differ over parks; x_{it} is the climate variable specific to each park and month; d_t is the monthly average for the US dollar index; $\bar{y}_t$ is the average of the proportional visitation (y) across all park units; $\bar{x}_t$ is the average of the climate variables across the park units for each month; and ε_{it} is the remaining unobservable disturbance. All estimated coefficients are represented as β'. Both $\bar{y}_t$ and $\bar{x}_t$ serve as proxy measures for unobserved common factors that may affect park visitation. Estimation of the panel time series models was completed using the xtcce command in Stata 14.0.

Results

Descriptive statistics

Average monthly values for each of the 10 climate variables and the visitation variable are provided in Table 2. To illustrate the presence of climate change within the parks, we note the historic averages (1901–2014), the averages for the years in which our analyses were preformed (1979–2014), and recent averages (2000–2014).

Zion National Park has and continues to be the most visited national park within the state, receiving an average of nearly 185,000 visitors a month since 1979. The past 15 years has seen average monthly visitation climb to over 222,000. Canyonlands National Park has received the fewest visits, averaging 27,500 visits a month since 1979. Visitation to the park has increased steadily, averaging over 35,000 between 2000 and 2014.

All of the parks have experienced warming temperatures since the early twentieth century, with average monthly mean temperatures rising between 0.6 °C (Arches and Canyonlands) and 1.1 °C (Bryce Canyon). Average monthly minimum temperatures have risen between 0.3 °C (Arches and Canyonlands) and 1.3 °C (Zion). Average monthly maximum temperatures have risen between 0.7 °C (Capitol Reef and Zion) and 0.9 °C (Canyonlands).

The landscapes of each park, excluding Canyonlands, have also become drier over the past 115 years. Mean monthly precipitation has declined between 0.4 mm per month (Arches) and 1.8 mm per month (Zion). The landscape of what is now Canyonlands National Park has actually received more precipitation in recent years, averaging 22.7 mm per month since 2000 (up from a historical average of 22.3 mm per month).

The correlations between each of the 10 climate variables and monthly visitation for each study park are illustrated in Figure 3. A positive correlation is noticeable for all of the temperature variables, with monthly average daily mean, minimum, and maximum temperatures showing the most distinct relationship with visitation; more variability can be seen in the correlations using both diurnal temperature range and potential evapotranspiration. Both of the precipitation variables (mm of precipitation per month and number of wet days per month) appear to be weakly related to monthly visitation, if at all. As might

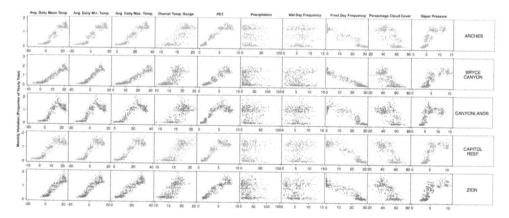

Figure 3. Correlations between monthly visitation (as a proportion of annual visitation) and each of the 10 climate variables for all five study parks. Source: Authors.

Table 2. Mean monthly values for each of the 10 climate variables and visitation for each study park.

	Arches National Park			Bryce Canyon National Park			Canyonlands National Park			Capitol Reef National Park			Zion National Park		
	1901–2014	1979–2014	2000–2014	1901–2014	1979–2014	2000–2014	1901–2014	1979–2014	2000–2014	1901–2014	1979–2014	2000–2014	1901–2014	1979–2014	2000–2014
Visitation															
Avg. monthly visitation	n/a	59,815	76,080	n/a	78,502	93,558	n/a	27,500	35,220	n/a	45,193	50,250	n/a	184,984	222,655
Temperature															
Monthly avg. daily mean temp.	9.6	10.2	10.2	8.1	8.7	9.2	9.8	10.4	10.4	8.5	9.0	9.3	9.1	9.6	10.1
Monthly avg. daily min. temp.	1.8	2.3	2.1	−0.1	0.6	1.1	1.8	2.3	2.1	0.4	1.0	1.3	0.7	1.3	2.0
Monthly avg. daily max. temp.	17.5	18.2	18.3	16.4	16.9	17.2	17.8	18.5	18.7	16.6	17.0	17.3	17.5	17.8	18.2
Diurnal temp. range	15.6	15.9	16.2	16.5	16.3	16.1	16.0	16.2	16.6	16.2	16.0	15.9	16.7	16.5	16.2
Potential evapotranspiration	3.8	3.9	3.8	3.9	3.9	4.0	3.9	3.9	3.9	3.9	3.9	3.9	4.2	4.2	4.2
Precipitation															
Precipitation	21.5	22.2	21.1	26.8	27.4	25.9	22.3	23.4	22.7	20.9	21.4	20.2	29.3	30.3	27.5
Wet day frequency	6.0	6.1	5.9	5.7	5.6	5.4	5.8	5.9	5.7	5.9	5.9	5.7	5.3	5.2	5.1
Other															
Frost day frequency	13.7	13.4	13.6	16.0	15.3	14.8	13.8	13.4	13.7	15.4	14.9	14.6	15.1	14.5	13.7
Percentage cloud cover	50.2	50.7	49.7	47.6	48.5	48.9	48.9	49.4	48.2	48.8	49.6	49.5	46.9	47.6	48.6
Vapor pressure	6.2	6.5	6.7	4.7	5.1	5.3	6.1	6.4	6.6	5.2	5.6	5.8	4.9	5.2	5.5

Source: Authors.

be expected, both the frequency of frost days and the percentage of cloud cover appear to be negatively related to visitation.

Common correlated effects estimation

Coefficients estimated for each of the 10 climate variables with the panel time series regression models are summarized in Table 3 (Columns 2–4). All four derivations of the temperature variable (monthly average daily mean temperature, monthly average daily minimum temperature, monthly daily average maximum temperature, and diurnal temperature range) were consistently and positively related to visitation (Coef. = 0.002–0.006; p = 0.002–0.004). Collectively, models including a temperature variable performed well, explaining between 56.2% (diurnal temperature range) and 82.1% (monthly average daily maximum temperature) of the variance in observed monthly visitation. Both of the precipitation variables (mm of precipitation per month and wet day frequency) were not significantly related to visitation (Coef. = 0.000–0.002; p = 0.425–0.476). When a precipitation variable was included in the model, almost none of the variance in monthly visitation was explained (R^2 < 0.001). Of the other climate variables explored, only potential evapotranspiration was significantly related to monthly visitation in the common correlated effects estimation (i.e. in the model with data from all five panels (parks) included). Potential evapotranspiration was positively and significantly related to monthly visitation (Coef. = 0.027; p = 0.005). This is logical, given the fact potential evapotranspiration is a derived measure of monthly and annual mean temperatures (Bonan, 2002). Frost day frequency, percentage cloud cover, and vapor pressure were not significantly related to monthly visitation (Coef. = −0.007–0.027; p = 0.129–0.383). However, the regressions including these variables as predictors did explain a good proportion of the variance in monthly visitation (R^2 = 0.507–0.854).

Panel-specific estimation

Looking beyond the common correlated effects estimations, we see more divergence in how the local climates of each individual park is related to that park's monthly visitation. Results from the panel-specific estimations are summarized in Table 3 (Columns 5–19); full model results are provided in Supplemental Table 2. The models suggest that temperature (monthly average daily means, minimums, and maximums) is positively and significantly related to monthly visitation for every park except Capitol Reef. Diurnal temperature range was also significantly related to monthly visitation at the two adjacent parks located in eastern Utah (Arches and Canyonlands), but not for the three other parks within the state. Canyonlands National Park appears to be an anomaly, with no observed relationship between temperature and visitation.

While precipitation, measured either via mm of precipitation per month or the number of wet days per month, was a very poor predictor of monthly visitation in the estimation using data from all five parks, the park-specific estimates suggest that it is related to visitation at some, but not all, of the parks within the state. The panel-specific models suggest precipitation (mm per month) is positively related to visitation at both Bryce Canyon and Capitol Reef National Parks (Coef. = −2.4e^{-4}–0.001; p < 0.016). The number of wet days per month was also positively related to visitation at Bryce Canyon and Canyonlands

Table 3. Estimated coefficients for each of the 10 climate variables derived via the common correlated effects estimations and the panel-specific estimations.

	Common correlated effects estimation			Arches National Park			Bryce Canyon National Park			Canyonlands National Park			Capitol Reef National Park			Zion National Park		
	Coef.	S.E.	Sig.	Coef.	S.E.	Sig.	Coef.	S.E.	Sig.	Coef.	S.E.	Sig.	Coef.	S.E.	Sig.	Coef.	S.E.	Sig.
Temperature																		
Monthly avg. daily mean temp.	0.006	0.002	**0.002**	0.004	0.000	**0.000**	0.012	0.001	**0.000**	0.010	0.002	**0.000**	0.002	0.002	0.253	0.003	0.001	**0.000**
Monthly avg. daily min. temp.	0.004	0.001	**0.007**	0.002	0.000	**0.000**	0.008	0.001	**0.000**	0.006	0.002	**0.000**	$4.9e^{-4}$	0.001	0.718	0.002	0.001	**0.004**
Monthly avg. daily max. temp	0.006	0.002	**0.000**	0.004	0.000	**0.000**	0.010	0.001	**0.000**	0.009	0.002	**0.000**	0.003	0.002	0.060	0.003	0.001	**0.000**
Diurnal temp. range	0.002	0.000	**0.000**	0.002	0.001	**0.002**	0.002	0.002	0.402	0.004	0.002	**0.041**	0.002	0.002	0.255	0.001	0.001	0.264
Potential evapotranspiration	0.027	0.009	**0.005**	0.020	0.003	**0.000**	0.017	0.008	**0.042**	0.063	0.008	**0.000**	0.010	0.008	0.200	0.023	0.004	**0.000**
Precipitation																		
Precipitation	$1.1e^{-4}$	$1.4e^{-4}$	0.425	$-8.6e^{-5}$	$4.9e^{-5}$	0.080	0.001	0.000	**0.000**	$2.6e^{-4}$	$1.6e^{-4}$	0.096	$-2.4e^{-4}$	$9.9e^{-4}$	**0.016**	$6.4e^{-5}$	$6.0e^{-5}$	0.287
Wet day frequency	0.002	0.003	0.476	0.001	0.000	0.112	0.010	0.001	**0.000**	0.005	0.001	**0.001**	-0.006	0.003	0.029	0.001	0.001	0.147
Other																		
Frost day frequency	-0.004	0.003	0.223	$-4.4e^{-4}$	$2.9e^{-4}$	0.127	-0.010	0.001	**0.000**	-0.012	0.001	**0.000**	-0.006	0.001	**0.000**	-0.004	$4.7e^{-4}$	**0.000**
Percentage cloud cover	0.002	0.001	0.129	$2.8e^{-5}$	$2.6e^{-4}$	0.915	0.006	0.001	**0.000**	0.001	0.001	0.472	0.002	0.001	0.117	$2.1e^{-4}$	$3.7e^{-4}$	0.565
Vapor pressure	-0.007	0.008	0.383	0.005	0.002	**0.002**	-0.018	0.004	**0.000**	-0.029	0.005	**0.000**	0.016	0.008	**0.042**	-0.009	0.002	**0.000**

Note: Estimated coefficients shown here were derived from 60 panel time series regression models (the 10 common correlated effects estimations plus 50 panel-specific estimations (10 climate variables × 5 panels). Not shown are: the estimates associated with the US dollar index; the estimates associated with the cross-sectional averages; and the constants. Full model results for each estimation are provided in Supplemental Table 1. Bold values are significant at the 0.05 level. Source: Authors.

(Coef. = 0.005–0.010; $p < 0.001$). It appears that while precipitation is a poor predictor of visitation across park units, it can be a good predictor of visitation within select units.

Of the other climate variables explored, potential evapotranspiration was positively related to monthly visitation at all of the parks (Coef. = 0.010–0.063; $p < 0.063$) with the exception of Capitol Reef (Coef. = 0.010; $p = 0.200$). Both frost day frequency and percentage cloud cover are related to monthly visitation at some parks, even though these climate variables are not good universal predictors of monthly visitation. Frost day frequency was negatively and significantly related to monthly visitation at each study park (Coef. = −0.012–−0.004; $p < 0.001$), with the exception of Arches. Percentage of cloud cover was significantly and positively related to monthly visitation within Bryce Canyon (Coef. = 0.006; $p < 0.001$), but it was not related to visitation at any of the other national parks (Coef. = $2.8e^{-5}$–0.006; $p > 0.117$). The final variable explored in the panel-specific models, vapor pressure, was significantly related to monthly visitation at all five national parks (Coef. = −0.029–0.005; $p < 0.042$). Recall however, that this variable was not significantly related to monthly visitation when data from all five park units were used (Coef. = −0.007; $p - 0.383$). This apparent contradiction is the product of vapor pressure being positively related to monthly visitation at Arches and Capitol Reef and negatively related to monthly visitation at Bryce Canyon, Canyonlands, and Zion. Whether or not a climate variable is a useful predictor of visitation depends on the geographic scale of the analysis. Some predictors, like precipitation and vapor pressure, may be useful at local, park-specific scales, even though their effects on visitation are washed out at larger, regional or nationwide analyses.

Discussion

A wide-range of climate variables can affect visitation

We began this investigation with two primary objectives, the first of which was to determine how a broad set of climate variables affect visitation across a tourism system. The majority of previous research on the relationship between climate and visitation has relied heavily on mean temperature as the primary factor affecting tourists' travel decisions; this is especially true of analyses of national and/or international travel patterns given the ubiquitous availability of temperature data (Bigano et al., 2006). Across this research, mean temperatures are almost always positively associated with visitation (e.g. Fisichelli et al., 2015; Hewer et al., 2016), suggesting a continued warming in mean air temperatures will contribute additional visitation pressures on tourism destinations. This is in addition to the pressures associated with growing populations and other potential drivers (White et al., 2016). The analysis presented here aligns with previous research exploring temperature as a driving factor shaping tourists' travel decisions. Across our study parks, we found a consistent positive relationship between the monthly average daily mean temperature and visitation levels. With the exception of Capitol Reef National Park, this relationship was highly significant ($p < 0.001$). Across all five parks, the use of monthly average daily mean temperatures explained a large proportion of the variance in monthly visitation rates ($R^2 \geq 0.814$).

While the monthly average of daily mean temperatures was a very good predictor of visitation to our study parks, it was not the best predictor. The monthly average of the daily maximum temperatures explained slightly more of the variance in visitation across the five parks ($R^2 \geq 0.821$). The reason for this may be attributable to the direct role daily maximum temperatures play in tourists' travel decisions. The climate averages of destinations are almost always presented through an average of daily maximum temperatures (as well as an average of daily minimum temperatures). When a tourist is considering different possible destinations, they rarely think about what the average daily temperature will be at each destination. Rather, individuals plan their travel behavior around the temperatures during the times when they will be most active; for most, this is mid-day or early afternoon when temperatures are at their highest. Future research may find a stronger behavioral response to average maximum daily temperatures relative to average mean daily temperatures, as our investigation did. We would expect future research to find similar patterns, especially if the analysis if focused more specifically on peak summer season visitation, which is more sensitive to temperature variations (Falk, 2014, 2015). Our analysis was conducted across all 12 months of the year, which likely masks some of the predictive power of the maximum temperature variable.

In addition to providing a more direct connection between climate and tourists' travel behavior, the average of maximum daily temperatures can illustrate the presence of a negative relationship between temperature and visitation. Previous research has suggested that the presence of a positive relationship between temperature and visitation rates only holds up to temperatures between 25 °C and 33 °C (Fisichelli et al., 2015; Hewer et al., 2016). Beyond this, visitation declines because it can become uncomfortable to be outside, especially at locations where many visitors are highly physically active (such as national parks). We observed this inverse u-shape relationship between temperature and monthly visitation for three of our study parks. At both Capitol Reef and Canyonlands National Parks, monthly visitation declines beyond an average maximum daily temperature between 25 °C and 33 °C. At Arches, visitation tends to level off once maximum daily temperatures hit this threshold. These patterns can clearly be seen in Figure 3. This inverse u-shape relationship was not noticeable in the data from either Bryce Canyon or Zion National Park, suggesting the relationship is not universal and that there may be confounding mitigating factors present. Both Bryce Canyon and Zion National Parks are characterized by seasonal streams and steep canyons (as well as slot canyons in Zion); both of these characteristics result in milder conditions than the actual temperature suggests. The unique geophysical characteristics of these parks (as well as others like them), may allow visitation rates to continue rising even with maximum daily temperatures well above 25 °C. These findings have both theoretical and managerial implications.

In regards to theory, it well understood that tourists are more capable of adapting to climate change than tourism destinations (Scott et al., 2008). Tourists can easily avoid destinations impacted by climate change by altering the timing or destination of their travel. Large tourism destinations however, especially those with large capital investments and immobile assets like national parks, have less adaptive capacity. The adaptive capacity of large tourism destinations is highly variable. Mather, Viner, and Todd (2005) suggest this variability is a function of a variety of factors including the physical environment and the topographical characteristics of specific destinations. Our data offer empirical support for this suggestion as it appears the seasonal streams and steep canyons which characterize the outdoor recreation opportunities within Bryce Canyon and Zion make those national parks more capable of adapting to a warmer and drier climate. This is one small empirical example of how large tourism destinations have highly variable adaptive capacities. Future research is needed to explore other factors which might affect this adaptive capacity (e.g. the nature of the tourism markets being served, the types of tourism facilities and attractions offered, etc.) (Mather et al., 2005).

As for management, knowing the unique geophysical characteristics of some parks may allow for more visitation, even with maximum daily temperatures well above 25 °C, may create additional challenges for destinations like Bryce Canyon and Zion which are already struggling to maintain their recreational infrastructure under extremely large visitation pressures (e.g. Zion National Park alone received 4.3 million visits in 2016). Zion National Park, for example, receives over 400,000 visitors a month in the peak summer months of July and August (National Park Service, 2017) and also has a $70 million maintenance backlog. Destinations like Zion and Bryce Canyon, whose unique geophysical characteristics and outdoor recreation opportunities allow visitors to feel cooler even as temperatures rise, are not likely to see a plateauing of summer visitation in the near future. Destinations like this will continue to face the daunting management challenge of accommodating more and more visitors with stagnant operational budgets. One potential solution is for the state's tourism campaigns to shift attention from the national parks of Utah to other attractive destinations and activities like the state's outstanding skiing attractions. The Utah Office of Tourism has recently began to head in this direction. In 2016, they launched 'The Road to Mighty' ad campaign (Utah Office of Tourism, 2017). The campaign highlights the many other attractions in Utah such as its state parks and national monuments. However, the popularity of Utah's national parks does not seem to be going away anytime soon, and neither are the challenges of managing park resources and visitors.

Aside from the temperature variables already discussed, our analysis explored the relationship between a wider array of climate variables and visitation than has been common in previous research. Principal among these were a set of precipitation variables, mm of precipitation per month and the number of wet days per month. By and large, these variables were very poor predictors of visitation. However, our panel-specific time series regression for Bryce Canyon National Park did reveal a positive and significant relationship between both precipitation variables and monthly visitation. Initially, this result may seem counterintuitive as one would assume wetter weather would either drive visitors away from the park or make it more likely for them to stay inside at their hotels outside of the park. Bryce Canyon is the only park within our sample that offers snow-based outdoor recreation opportunities, including groomed trails for cross-country skiing and snowshoeing. Data from Bryce Canyon illustrate consistent visitation in the winter months and, as illustrated in our results, visitation that increases when more snow is present. Future investigations into the relationship between climate and visitation should not disregard the possibility of precipitation being a significant factor shaping visitors' travel decisions. Smaller-scale (e.g. park specific) and activity-specific (e.g. beach recreation, skiing/snowboarding, etc.) research is much more likely to find a behavioral response to precipitation. For example, recent research has found that extreme dry conditions can lead to notable declines in visitation to mountainous national parks in the western US (Jedd et al., 2017).

One of the limitations of our investigation is the inability to examine whether the relationships observed between climatic conditions and visitation are similar for both local visitors and distant visitors. Previous research suggests climate change affects the travel behaviors of international and domestic visitors differently, with international visitors being more likely to visit a destination regardless of the weather conditions during their travel (Scott, Gössling, & Hall, 2012). Domestic visitors, especially those living closer to the destination, are able to alter the timing or duration of their visit more easily than international visitors. Previous research does not provide much insight beyond these logical and intuitive differences. More focused research is needed, especially for prominent tourism destinations like some US National Parks, which receive a substantial amount of visitation from both domestic and international markets.

Scale matters in analyses of climate and visitation

The second objective of this investigation was to identify which climate variables affect visitation across an entire tourism system and which climate variables only affect visitation at specific destinations. By conducting our analyses at two distinct scales, across a regional recreation system consisting of five national parks and at each of those national parks individually, we have been able to ascertain whether there is a difference between the individual climate variables that are regionally-significant drivers of visitation and those that are locally-significant drivers of visitation. Our analyses illustrate that scale matters in analysis of the relationship between climatic conditions and visitation patterns. All of the temperature variables we examined were good predictors of visitation at both a regional scale and the scale of most individual park units. However, for one park (Capitol Reef) the relationship between temperature and visitation was not statistically significant. This suggests that the positive relationship between temperature and visitation, while being commonly found in mid-latitude tourism destinations, is not a universal maxim. Many destinations, like Capitol Reef, may have visitation trends that plateau in the summer months due to a variety of factors. The landscape of Capitol Reef is a high desert with little vegetative cover and few streams. These geophysical conditions can make the hot summer months uncomfortable, resulting in a monthly visitation profile that is characterized by stagnant visitation between the months of May and September. This finding is noteworthy and highlights the need for regional climate-specific tourism policy and resource management approaches to not assume that temperature and visitation are always positively related. Careful consideration needs to be given to local geophysical, institutional, and social factors that may mediate the relationship between temperature and visitation. In the same vein, our analyses also revealed other climate variables that were poor predictors of visitation across the region, but that were highly significant predictors of visitation at the scale of individual parks. Precipitation (mm of precipitation per month), frost day frequency, and vapor pressure were not significantly related to monthly visitation in our regional model (i.e. using data from all five parks). However, they were significantly related to visitation at individual park units. This finding suggests the need to cast a wide net in considering the specific climatic conditions that influence visitation across a recreation system. Collectively, these findings highlight the need for future research, climate-related policy formation, and resource management planning to be vigilant of the fact that scale matters in the relationship between climate and visitation.

Conclusion

By exploring how a broad set of climate variables affect visitation across a tourism system, our analysis has demonstrated that different climate variables affect visitation to managed natural areas in different ways. The relationship between specific climate variables and visitation is a function of the geophysical characteristics and recreational opportunities that are available at specific destinations. Climate is obviously related to visitation patterns, but how and why it is related to visitation patterns is a product of a more diverse array of factors (e.g. vegetative cover, geographic relief, types of recreational activities supported, etc.) than are commonly considered in research on climate and tourism.

By identifying which climate variables affect visitation across an entire recreation system and which climate variables only affect visitation at specific destinations, our analyses demonstrated how the spatial scale of an analysis yields quantifiably different results. Ignoring the issue of spatial scale in climate-focused visitation research can result in misinformed climate-related policy and in poorly developed resource management frameworks. Future investigations need to be vigilant in considering how the scope of their analysis informs the inferences that can be made, and consequently, the types of policies and management decisions that can be recommended by their analyses.

Notes

1. We are working from the definition of a tourism system as the combination of tourist generating regions, the tourist destination region, and the transit region. Our study is focused specifically on the tourist destination region, defined as the 'locations which attract tourists to stay temporarily, and in particular those features which inherently contribute to that attraction' (Leiper, 1979, p. 397).
2. Panel data describe a sample of units (in this case, national parks) which are repeatedly measured over time (in this case, each month from January 1979 to December 2014). When panel data describe a relatively few number of units over a long time period (small N, large t) they are described as a macro panel (Hsiao, 2014).

Acknowledgement

This research was supported in part by a grant from Utah State University Extension.

Disclosure statement

No potential conflict of interest was reported by the authors.

Funding

Utah State University Extension.

ORCID

Jordan W. Smith (iD) http://orcid.org/0000-0001-7036-4887
Emily Wilkins (iD) http://orcid.org/0000-0003-3055-4808

References

Amelung, B., & Nicholls, S. (2014). Implications of climate change for tourism in Australia. *Tourism Management, 41*, 228–244. doi:10.1016/j.tourman.2013.10.002

Amelung, B., & Viner, D. (2006). Mediterranean tourism: Exploring the future with the tourism climatic index. *Journal of Sustainable Tourism, 14*(4), 349–366. doi:10.2167/jost549.0

Anastasopoulos, P. G. (1989). The U.S. travel account: The impact of fluctuations on the U.S. dollar. *Hospitality Education and Research Journal, 13*(3), 469–481. doi:10.1177/109634808901300349

Banerjee, A., & Carrion-i-Silvestre, J. L. (2017). Testing for panel cointegration using common corre- lated effects estimators. *Journal of Time Series Analysis, 38*(4), 610–636. doi:10.1111/jtsa.12234

Barrios, S., & Ibañez, J. N. (2015). Time is of the essence: Adaptation of tourism demand to climate change in Europe. *Climatic Change, 132*(4), 645–660. doi:10.1007/s10584-015-1431-1

Beaudin, L., & Huang, J.-C. (2014). Weather conditions and outdoor recreation: A study of New Eng- land ski areas. *Ecological Economics, 106*, 56–68.

Becken, S. (2013). Measuring the effect of weather on tourism: A destination-and activity-based anal- ysis. *Journal of Travel Research, 52*(2), 156–167.

Becken, S., & Wilson, J. (2013). The impacts of weather on tourist travel. *Tourism Geographies, 15*(4), 620–639.

Berrittella, M., Bigano, A., Roson, R., & Tol, R. S. J. (2006). A general equilibrium analysis of climate change impacts on tourism. *Tourism Management, 27*(5), 913–924.

Bigano, A., Hamilton, J. M., Maddison, D. J., & Tol, R. S. (2006). Predicting tourism flows under climate change. *Climatic Change, 79*(3), 175–180.

Bonan, G. (2002). *Ecological climatology*. Cambridge: CUP.

Burdisso, T., & Sangiácomo, M. (2016). Panel time series: Review of the methodological evolution. *The Stata Journal, 16*(2), 424–442.

Coakley, J., Fuertes, A.-M., & Smith, R. (2002). A principal components approach to cross-section dependence. *Paper presented at the 10th International Conference on Panel Data*. Berlin, Germany.

Coombes, E. G., Jones, A. P., & Sutherland, W. J. (2009). The implications of climate change on coastal visitor numbers: A regional analysis. *Journal of Coastal Research, 25*(4), 981–990.

Dawson, J., Scott, D., & Havitz, M. (2013). Skier demand and behavioural adaptation to climate change in the US Northeast. *Leisure, 37*(2), 127–143.

Denstadli, J. M., Jacobsen, J. K. S., & Lohmann, M. (2011). Tourist perceptions of summer weather in Scandanavia. *Annals of Tourism Research, 38*(3), 920–940.

Falk, M. (2014). Impact of weather conditions on tourism demand in the peak summer season over the last 50 years. *Tourism Management Perspectives, 9*, 24–35.

Falk, M. (2015). Summer weather conditions and tourism flows in urban and rural destinations. *Cli- matic Change, 130*(2), 201–222. doi:10.1007/s10584-015-1349-7

Fisichelli, N. A., Schuurman, G. W., Monahan, W. B., & Ziesler, P. S. (2015). Protected area tourism in a changing climate: Will visitation at US national parks warm up or overheat? *PLoS ONE, 10*(6), e0128226. doi:10.1371/journal.pone.0128226

Fisichelli, N. A., & Ziesler, P. S. (2015a). *Arches National Park: How might future warming alter visitation?* *(Park visitation and climate change: Park-specific brief)*. Washington, DC: National Park Service. Retrieved from https://irma.nps.gov/DataStore/Reference/Profile/2222467

Fisichelli, N. A., & Ziesler, P. S. (2015b). *Bryce Canyon National Park: How might future warming alter visitation?* *(Park visitation and climate change: Park-specific brief)* . Washington, DC: National Park Service. Retrieved from https://irma.nps.gov/DataStore/Reference/Profile/2222778

Fisichelli, N. A., & Ziesler, P. S. (2015c). *Canyonlands National Park: How might future warming alter visitation?* *(Park visitation and climate change: Park-specific brief)* . Washington, DC: National Park Service. Retrieved from https://irma.nps.gov/DataStore/Reference/Profile/2222790

Fisichelli, N. A., & Ziesler, P. S. (2015d). *Capitol Reef National Park: How might future warming alter visitation?* *(Park visitation and climate change: Park-specific brief)* . Washington, DC: National Park Service. Retrieved from https://irma.nps.gov/DataStore/Reference/Profile/2222791

Fisichelli, N. A., & Ziesler, P. S. (2015e). *Zion National Park: How might future warming alter visitation?* *(Park visitation and climate change: Park-specific brief)*. Washington, DC: National Park Service. Retrieved from https://irma.nps.gov/DataStore/Reference/Profile/2222766

Førland, E. J., Jacobsen, J. K. S., Denstadli, J. M., Lohmann, M., Hanssen-Bauer, I., Hygen, H. O., & Tømmervik, H. (2013). Cool weather tourism under global warming: Comparing Arctic summer tourists' weather preferences with regional climate statistics and projections. *Tourism Management, 36*, 567–579.

Gössling, S., Abegg, B., & Steiger, R. (2016). "It was raining all the time!": Ex post tourist weather perceptions. *Atmosphere, 7*(1), 10. doi:10.3390/atmos7010010

Gössling, S., & Hall, C. M. (2006). Uncertainties in predicting tourist flows under scenarios of climate change. *Climatic Change, 79*(3), 163–173. doi:10.1007/s10584-006-9081-y

Hamilton, J. M., & Lau, M. A. (2006). The role of climate information in tourist destination choice decision making. In S. Gössling & C. M. Hall (Eds.), *Tourism and global environmental change: Ecological, economic, social and political interrelationships* (Vol. 229, pp. 229–250). New York, NY: Routledge.

Hansen, A. J., Davis, C. R., Piekielek, N., Gross, J., Theobald, D. M., Goetz, S., ... DeFries, R. (2011). Delineating the ecosystems containing protected areas for monitoring and management. *BioScience, 61*(5), 363–373. doi:10.1525/bio.2011.61.5.5

Harris, I., Jones, P. D., Osborn, T. J., & Lister, D. H. (2014). Updated high-resolution grids of monthly climatic observations – The CRU TS3.10 Dataset. *International Journal of Climatology, 34*(3), 623–642. doi:10.1002/joc.3711

Hewer, M., Scott, D., & Fenech, A. (2016). Seasonal weather sensitivity, temperature thresholds, and climate change impacts for park visitation. *Tourism Geographies, 18*(3), 297–321. Retrieved from https://doi.org/10.1080/14616688.2016.1172662

Hewer, M., Scott, D., & Gough, W. A. (2015). Tourism climatology for camping: A case study of two Ontario parks (Canada). *Theoretical and Applied Climatology, 121*(3), 401–411. doi:10.1007/s00704-014-1228-6

Hsiao, C. (2014). *Analysis of panel data* (3rd ed.). Cambridge: Cambridge University Press.

Jedd, T. M., Hayes, M. J., Carrillo, C. M., Haigh, T., Chizinski, C., & Swigart, J. (2017). Measuring park visitation vulnerability to climate extremes in U.S. Rockies National Parks tourism. *Tourism Geographies*. Advance online publication. doi:10.1080/14616688.2017.1377283

Jones, B., & Scott, D. (2006). Climate change, seasonality and visitation to Canada's national parks. *Journal of Park & Recreation Administration, 24*(2).42–62.

Kapetanios, G., Pesaran, M. H., & Yamagata, T. (2011). Panels with non-stationary multifactor error structures. *Journal of Econometrics, 160*(2), 326–348. Retrieved from https://doi.org/10.1016/j.jeconom.2010.10.001

Köberl, J., Prettenthaler, F., & Bird, D. N. (2016). Modelling climate change impacts on tourism demand: A comparative study from Sardinia (Italy) and Cap Bon (Tunisia). *The Science of the Total Environment, 543*(Pt B), 1039–1053. doi:10.1016/j.scitotenv.2015.03.099

Leiper, N. (1979). The framework of tourism: Towards a definition of tourism, tourist, and the tourist industry. *Annals of Tourism Research, 6*(4), 390–407. doi:10.1016/0160-7383(79)90003-3

Lise, W., & Tol, R. S. J. (2002). Impact of climate on tourist demand. *Climatic Change, 55*(4), 429–449. doi:10.1023/A:1020728021446

Liu, T.-M. (2016). The influence of climate change on tourism demand in Taiwan national parks. *Tourism Management Perspectives, 20*, 269–275. doi:10.1016/j.tmp.2016.10.006

Loomis, J. B., & Richardson, R. B. (2006). An external validity test of intended behavior: Comparing revealed preferences and intended visitation in response to climate change. *Journal of Environmental Planning and Management, 49*(4), 621–630.

Mather, S., Viner, D., & Todd, G. (2005). Climate and policy changes: Their implications for international tourism flows. In C. M. Hall & J. Higham (Eds.), *Tourism, Recreation and Climate Change* (pp. 63–85). Clevedon: Channel View Publications.

Mieczkowski, Z. (1985). The tourism climatic index: A method of evaluating world climates for tourism. *Canadian Geographer / Le Géographe Canadien, 29*(3), 220–233. doi:10.1111/j.1541-0064.1985.tb00365.x

Monahan, W. B., & Fisichelli, N. A. (2014). Climate exposure of US national parks in a new era of change. *PLoS ONE, 9*(7), e101302. doi:10.1371/journal.pone.0101302

Moreno, A., & Amelung, B. (2009). Climate change and tourist comfort on Europe's beaches in summer: A reassessment. *Coastal Management, 37*(6), 550–568. doi:10.1080/08920750903054997

National Park Service. (2017). Integrated resource management applications. Retrieved from https://irma.nps.gov/Portal/

NPS Natural Resource Inventory and Monitoring Division. (2016). Area of analysis source polygons – NPS boundary-derived (park, 3km, and 30km) – Landscape dynamics project. Retrieved from https://irma.nps.gov/DataStore/Reference/Profile/2235933

Patrolia, E., Thompson, R., Dalton, T., & Hoagland, P. (2017). The influence of weather on the recreational uses of coastal lagoons in Rhode Island, USA. *Marine Policy, 83*(Supplement C), 252–258. doi:10.1016/j.marpol.2017.06.019

Perch-Nielsen, S., Amelung, B., & Knutti, R. (2010). Future climate resources for tourism in Europe based on the daily Tourism Climatic Index. *Climatic Change, 103*(3–4), 363–381. doi:10.1007/s10584-009-9772-2

Pesaran, M. H. (2006). Estimation and inference in large heterogeneous panels with a multifactor error structure. *Econometrica, 74*(4), 967–1012. doi:10.1111/j.1468-0262.2006.00692.x

Richardson, R. B., & Loomis, J. B. (2004). Adaptive recreation planning and climate change: A contingent visitation approach. *Ecological Economics, 50*(1), 83–99. doi:10.1016/j.ecolecon.2004.02.010

Rosselló-Nadal, J. (2014). How to evaluate the effects of climate change on tourism. *Tourism Management, 42*(Supplement C), 334–340. doi:10.1016/j.tourman.2013.11.006

Rosselló-Nadal, J., Riera-Font, A., & Cárdenas, V. (2011). The impact of weather variability on British outbound flows. *Climatic Change, 105*(1), 281–292. doi:10.1007/s10584-010-9873-y

Rutty, M., & Andrey, J. (2014). Weather forecast use for winter recreation. *Weather, Climate, and Society, 6*(3), 293–306.

Rutty, M., & Scott, D. (2010). Will the Mediterranean become "too hot" for tourism? A reassessment. *Tourism and Hospitality Planning & Development, 7*(3), 267–281. doi:10.1080/1479053X.2010.502386

Rutty, M., & Scott, D. (2014). Thermal range of coastal tourism resort microclimates. *Tourism Geographies, 16*(3), 346–363. doi:10.1080/14616688.2014.932833

Rutty, M., & Scott, D. (2016). Comparison of climate preferences for domestic and international beach holidays: A case study of Canadian travelers. *Atmosphere, 7*(2), 30. doi:10.3390/atmos7020030

Rutty, M., Scott, D., Johnson, P., Jover, E., Pons, M., & Steiger, R. (2015). Behavioural adaptation of skiers to climatic variability and change in Ontario, Canada. *Journal of Outdoor Recreation and Tourism, 11*, 13–21. doi:10.1016/j.jort.2015.07.002

Scott, D., Amelung, B., Becken, S., Ceron, J.-P., Dubois, G., Gössling, S., ... Simpson, M. C. (2008). *Climate change and tourism: Responding to global challenges*. Madrid: World Tourism Organization and United Nations Environment Programme.

Scott, D., Dawson, J., & Jones, B. (2008). Climate change vulnerability of the US Northeast winter recreation-tourism sector. *Mitigation and Adaptation Strategies for Global Change, 13*(5), 577–596. doi:10.1007/s11027-007-9136-z

Scott, D., Gössling, S., & de Freitas, C. R. (2008). Preferred climates for tourism: Case studies from Canada, New Zealand and Sweden. *Climate Research, 38*(1), 61–73. doi:10.3354/cr00774

Scott, D., Gössling, S., & Hall, C. M. (2012). International tourism and climate change. *WIREs Climate Change, 3*, 213–232.

Scott, D., & Jones, B. (2007). A regional comparison of the implications of climate change for the golf industry in Canada. *The Canadian Geographer / Le Géographe Canadien, 51*(2), 219–232. doi:10.1111/j.1541-0064.2007.00175.x

Scott, D., Jones, B., & Konopek, J. (2007). Implications of climate and environmental change for nature-based tourism in the Canadian Rocky Mountains: A case study of Waterton Lakes National Park. *Tourism Management, 28*(2), 570–579. doi:10.1016/j.tourman.2006.04.020

Scott, D., McBoyle, G., & Schwartzentruber, M. (2004). Climate change and the distribution of climatic resources for tourism in North America. *Climate Research, 27*, 105–117.

Scott, D., Rutty, M., Amelung, B., & Tang, M. (2016). An inter-comparison of the holiday climate index (HCI) and the tourism climate index (TCI) in Europe. *Atmosphere, 7*(6), 80. doi:10.3390/atmos7060080

Serquet, G., & Rebetez, M. (2011). Relationship between tourism demand in the Swiss Alps and hot summer air temperatures associated with climate change. *Climatic Change, 108*(1), 291–300. doi:10.1007/s10584-010-0012-6

Smith, J. W., Seekamp, E., McCreary, A., Davenport, M., Kanazawa, M., Holmberg, K., ... Nieber, J. (2016). Shifting demand for winter outdoor recreation along the North Shore of Lake Superior under variable rates of climate change: A finite-mixture modeling approach. *Ecological Economics, 123*, 1–13.

Steiger, R., Abegg, B., & Jänicke, L. (2016). Rain, rain, go away, come again another day. Weather preferences of summer tourists in mountain environments. *Atmosphere, 7*(5), 63. doi:10.3390/atmos7050063

U.S. Travel Association. (2016 , November 2). Study: More overseas visitors choosing U.S. national parks [Text]. Retrieved from https://www.ustravel.org/press/study-more-overseas-visitors-choosing-us-national-parks

Utah Office of Tourism. (2017). Road to mighty. Retrieved from https://www.visitutah.com/road-to-mighty/

White, E. M., Bowker, J. M., Askew, A. E., Langner, L. L., Arnold, J. R., & English, D. B. K. (2016). *Federal outdoor recreation trends: Effects on economic opportunities* (General Technical Report No. PNW-GTR-945). Olympia, WA: Pacific Northwest Research Station.

Yu, G., Schwartz, Z., & Walsh, J. E. (2009). A weather-resolving index for assessing the impact of climate change on tourism related climate resources. *Climatic Change, 95*(3–4), 551–573. doi:10.1007/s10584-009-9565-7

Weather sensitivity and climate change perceptions of tourists: a segmentation analysis

Emily Wilkins ⓘD, Sandra de Urioste-Stone ⓘD, Aaron Weiskittel and Todd Gabe

ABSTRACT

Many communities rely on tourism spending, so it is important to understand any potential changes to tourist flows resulting from changing climate and weather patterns. However, tourists are not a homogenous group, as they have different motivations, values, and goals. Therefore, the purpose of this investigation is to better understand potentially varying perceptions and behavior of different tourist types, specifically in regards to their weather sensitivity, climate change concern, and behavioral intention for climate change mitigation. Tourists were randomly surveyed at 20 locations throughout the state of Maine in the United States ($n = 704$). Segmentation analysis on the activities tourists participated in yielded three segments of Maine tourists: non-nature-based tourists (50.6%), nature-based generalists (16.2%), and nature-based specialists (33.2%). Differences across segments were explored for perceptions of weather impacts, climate change concern, and mitigation intent. Additionally, weather sensitivity was analyzed based on type of overnight accommodations to better understand if this also had a role in differences. Non-nature-based tourists thought that weather variables were less influential during their travels in Maine than the other segments, while nature-based generalists perceived weather to have the highest influence. Additionally, nature-based specialists had the highest level of climate change belief, while nature-based generalists had the highest willingness to engage in climate change mitigation behavior. Results are useful to understand how segments of tourism demand may be altered with a changing climate, such as increased temperatures, precipitation, and storms, and what groups may be the most beneficial to target for marketing or educational efforts to reduce the impact of climate change.

摘要

很多社区依赖旅游花费, 因此了解旅游流针对气候与天气变化而具有的潜在变化是很重要的。然而, 旅游者并不是一个同质化的群体, 他们有不同的旅游动机、价值观与旅游目的。因此, 本文的目的是深入理解不同类型旅游者对天气变化敏感度、气候变化关注度以及缓解气候变化的行为意图的潜在的多样化的认知与行为。本次调查对缅因州20个地点的旅游者进行随机调查 (样本量为704)。本文对旅游者参与的活动进行聚类分析, 产生三种类型旅游者:非自然旅游者 (占50.6%) 、通才型自然旅游者 (占16.2%) 和专家型自然旅游者 (占33.2%) 。然后, 本文探讨了各个旅游者细分群体对气候影响、气候变化关注度与缓冲行为方面的认知差异。为了较好地理解过夜住宿对上述认知行为差异是否有影响, 我们根据旅游者的过夜住宿类型分析了他们对天气敏感度方面的差异, 发现非自然旅游者比其他类型旅游者认为, 天气变量在缅因州旅行中较不重要, 而通才型自然旅游者认为天气变量最为重要。另外, 专家型自然旅游者相信气候变化的程度最高, 而通才型自然旅游者有最强的意愿参与气候变化的缓解行为。本结果有助于理解不同旅游需求群体伴随诸如气温升高、降雨和风暴等气候变化而变化的过程, 以及如果为减少气候变化的影响而向旅游者进行宣传教育, 对哪种类型的旅游者宣传教育效果最好。

Introduction

Tourism is important for many economies globally, supporting an estimated 292 million jobs and accounting for 10.2% of GDP in 2016 (World Travel & Tourism Council, 2017). In 2016 in the United States, tourism supported 14.2 million jobs and comprised 8.1% of GDP (World Travel & Tourism Council, 2017). Therefore, it is important to understand any potential changes to tourism flows, as these changes have real economic and social impacts for communities. Although tourism is important globally, some communities are more reliant on tourists' spending than others, such as gateway towns into national parks and protected areas.

Tourism is one of the largest industries in the state of Maine, contributing $5.99 billion in direct expenditures in 2016 (Maine Office of Tourism, 2017, p. 19). Many communities in Maine rely on tourism to support jobs and the economy; in fact, 71% of residents believe tourism is the most important economic driver for the state (Maine Office of Tourism, 2016). Therefore, potential changes to tourism in Maine could impact residents' livelihoods. Since Maine is a heavily forested and rural state with both mountain and coastal assets, much of the tourism is nature-based. Although there are various factors that could impact tourism, such as the state of the economy, overall tourism trends, or tourism marketing, this study focuses on how weather and a changing climate could impact tourism and different tourist groups. Climate change is already impacting tourism globally (Gössling, Scott, Hall, Ceron, & Dubois, 2012), and tourists already perceive impacts of climate change to tourism in Maine (De Urioste-Stone, Scaccia, & Howe-Poteet, 2015). This study aims to better understand both weather sensitivities and climate change perceptions of different tourist groups to understand multiple facets of how tourism-dependent communities could be impacted under a changing climate.

Weather, climate change, and tourism

Tourists are highly influenced by weather and climate, since these impact destination selection, trip timing, and trip satisfaction (e.g. Becken & Hay, 2007). Weather affects tourism by influencing activities participated in, travel and transportation, and the length of visitors' stays (e.g. Denstadli, Jacobsen, & Lohmann, 2011; Smith, 1993). For example, average sunshine and temperature both positively impacted domestic overnight stays in

Austria during peak season, while average precipitation had a negative effect (Falk, 2014). In New Zealand, a survey found that 39% of international tourists changed their trip timing as a result of the weather, and 51% changed activities due to weather (Becken & Wilson, 2013). Additionally, many travel bloggers mention weather variables when recounting their trips (Jeuring & Peters, 2013). However, the effect of weather may be different depending on the location and the type of tourist, as one study showed that urban tourists in Hong Kong were minimally impacted by weather (McKercher, Shoval, Park, & Kahani, 2015).

Past research on weather and tourism tend to measure the influence of four core variables: air temperature, rain, sunshine, and wind (Hewer, Scott, & Gough, 2015; Rutty & Scott, 2010; Scott, Gössling, & de Freitas, 2008; Steiger, Abegg, & Jänicke, 2016). The perceived importance of weather varies depending on the destination. For example, temperature tends to matter more for urban tourism (Rutty & Scott, 2010; Scott et al., 2008), while sunshine and rain are more important for beach tourism (Moreno, Amelung, & Santamarta, 2009; Scott et al., 2008), and rain is the most influential for mountain tourism (Scott et al., 2008; Steiger et al., 2016). Many studies have concluded that wind has the least importance behind sunshine, rain, and temperature (Hewer et al., 2015; Moreno & Amelung, 2009; Rutty & Scott, 2010; Scott et al., 2008; Steiger et al., 2016).

Mieczkowski's tourism climate index (TCI; 1985) also used these four variables to create an index for desirability of tourism climates, with daytime/daily comfort (measured by temperature and humidity) having the highest impact (50%), followed by precipitation and sunshine (20% each), and wind (10%). Since the impact of weather does vary based on location, Morgan et al. (2000) created an index for beach tourism, where temperature had less of an impact, and precipitation was the most important weather variable. In addition to varying based on location and type of tourism, weather sensitivity may also vary based on overnight accommodation. A recent study by Hewer et al. (2015) found that the most influential weather conditions may be different for campers, with sunshine and temperature having the highest importance for camper satisfaction, but heavy rain and strong winds most likely to cause visitors to leave early.

Although weather is important during a vacation, overall climate, or the long-term average of weather, is influential in determining destination selection and when tourists visit (Becken & Hay, 2007). However, perceptions of what constitutes an acceptable climate differ between tourists from different areas or with varying backgrounds (e.g. Gómez Martín, 2005; Rutty & Scott, 2016; Scott et al., 2008). Nevertheless, despite these varying tourist perceptions, changing weather and climate likely does impact tourism flows globally by altering which destinations people perceive as attractive and in which season they travel (Becken, 2012; Gössling et al., 2012).

This is especially pertinent as climate is changing all around the world. From 1800–2012, average land and ocean temperatures have risen by 0.85 °C (Intergovernmental Panel on Climate Change, 2013). In addition, there has been an increase in the frequency and intensity of extreme weather events (Intergovernmental Panel on Climate Change, 2013). This is already impacting outdoor recreation and tourism globally by shifting where and when visitors decide to travel (Gössling et al., 2012). For example, a study in a Canadian national park found that when just modeling temperature and precipitation under climate change scenarios, visitation was expected to increase (Scott, Jones, & Konopek, 2007). However, when visitors were surveyed at the same park, many people said they would visit less often or not at all based on predicted environmental changes, such as wildlife population decline, fewer glaciers, and higher probabilities of wildfires (Scott et al., 2007). Additionally, based on the weather variables in the TCI and future climate models, attractiveness of destinations and seasonality is expected to shift across North America, with some destinations expected to have an increase in climate resources for tourism, while others a decrease (Scott, McBoyle, & Schwartzentruber, 2004).

Climate change and tourism visitation in Maine

Over the last century, the climate of Maine has changed by becoming warmer, wetter, and having more storm variability. The mean annual temperature has increased by 1.7 °C from 1895 to 2014, and mean annual precipitation has increased by 13% (Fernandez et al., 2015). This could already be impacting visitors' travels to and within Maine. Furthermore, climate change scenarios predict Maine will continue to become warmer and wetter. Using IPCC climate change scenarios, models predict that by the middle of the twenty-first century, average annual temperature in Maine will increase by an additional 1.1–1.7 °C, and precipitation will increase an additional 1%–7% from 2015 levels (Fernandez et al., 2015).

Additionally, climate change is predicted to have a large impact on winter tourism, since future projections show the increase in precipitation as more rain and less snow (Fernandez et al., 2015). Research suggests the snowmobile season in Maine will be reduced under climate change (Scott, Dawson, & Jones, 2008), and that only 57% of Maine alpine ski locations will be able to maintain a season length of at least 100 days by the 2050s under low emissions scenarios (Dawson & Scott, 2013). A changing climate means that weather is also changing, so it is important to understand how tourists perceive the impact of weather on their travels in order to better understand how visitation in Maine could change in the future.

Recent studies in Maine showed that tourists are already noting the impact climate change could have on tourism (De Urioste-Stone et al., 2015), and many of them would change their future visitation based on possible changing weather conditions (De Urioste-Stone, Le, Scaccia, & Wilkins, 2016). Of summer visitors to Mount Desert Island, Maine, 61% believed climate change would affect tourism in the area, with the majority thinking it would have a negative impact. Visitors expressed concern over an increase in extreme temperatures, the increased frequency of rain and storm events, and sea level rise (De Urioste-Stone et al., 2015). Understanding visitors' perceptions is important because perceptions influence behavior (e.g. Denstadli et al., 2011). Additionally, visitors' intended visitation response to climate change scenarios in Rocky Mountain National Park was not significantly different than revealed preferences from regression models (Loomis & Richardson, 2006). Therefore, investigating tourists' perceptions of their behavior can be useful to understand actual behavior. This is important so that communities and protected areas can prepare ahead for changes in spending and visitation patterns.

Tourism and environmental engagement

Tourism and climate change have a multifaceted relationship because tourism is a contributor to anthropogenic climate change, but climate change impacts tourists and tourism destinations as well (e.g. Scott, Gössling, & Hall, 2012). The impacts of tourism include increased energy, emissions, food, and water (Gössling & Peeters, 2015). Additionally, much research has noted the high carbon footprint of tourism (e.g. Cadarso, Gómez, López, Tobarra, & Zafrilla, 2015; Sharp, Grundius, & Heinonen, 2016).

As tourism demand continues to grow, and thus emissions from tourism likely continue increasing, it will be important to understand how to mitigate some of the negative consequences of tourism. A study by McKercher, Prideaux, Cheung, and Law (2010) found that tourists are largely not willing to change their travel behavior to reduce their carbon emissions. Additionally, previous studies have evaluated tourists' willingness to pay for carbon offsets while flying (e.g. Choi, Ritchie, & Fielding, 2016; Gössling, Haglund, Kallgren, Revahl, & Hultman, 2009; Segerstedt & Grote, 2016). Although few travellers participated in airline carbon-offsetting programs, this could be due to a lack of knowledge; more travellers indicated a willingness to pay than those who actually bought carbon offsets (Gössling et al., 2009).

Understanding beliefs and perceptions on climate change is especially important to influence environmental behavior because perceptions influence policies and decision-making (Brownlee, Hallo, & Krohn, 2013; Brownlee, Powell, & Hallo, 2013). Additionally, general beliefs and concern for an issue (such as climate change) are important to measure since belief and concern are a precursor for action (Roser-Renouf, Maibach, Leiserowitz, & Zhao, 2014).

Coupling the importance of climate change perceptions with the known weather sensitivity of tourism, the objectives of this study are to (1) explore perceptions of how weather affects different tourist groups to better understand how behavior might differ under future climate change conditions and (2) examine climate change concern and willingness to take action across tourist groups. In this study, tourists are defined as temporary visitors who stay at least 24 hours away from their permanent residence (Leiper, 1979). In recognition that tourists are not a homogenous group (Wight, 2001), this study utilizes segmentation analysis to compare and contrast differences among tourist groups in Maine.

Methodology

Study site

Maine is located in the northeastern part of the United States and is the most forested state in the USA, with about 90% of land cover being forests (Forests for Maine's Future, 2011). Tourism, along with forest products, is one of the largest industries in the state. Maine has eight tourism regions, which offer a wide array of tourism opportunities, ranging from beaches to mountains to urban tourism (Explore Maine, 2014). In 2016, Maine had a total of 18.9 million tourism-related overnight visits, most of which were from out-of-state. Summer is the most popular season for tourism, as 51.3% of overnight tourists visited in the summer (Maine Office of Tourism, 2017).

This study was conducted at 20 locations across Maine, including visitor's centers, state parks, Bangor International Airport (BIA), a chamber of commerce, and Acadia National Park (Figure 1). The large number of sampling sites was chosen because a previous study

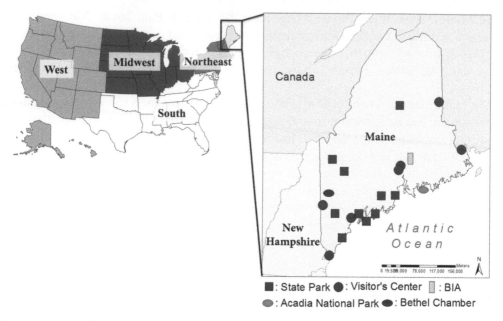

■ : State Park ● : Visitor's Center ▯ : BIA
◑ : Acadia National Park ◖ : Bethel Chamber

Figure 1. A map of the sampling sites throughout Maine.

that undertook segmentation analysis of tourists in Norway suggested future research should survey visitors at a greater number of places and at diverse locations to avoid bias (Mehmetoglu, 2007).

Survey design and sampling procedure

This study consisted of an on-site intercept questionnaire, followed by a longer self-administered online questionnaire. Participants of the on-site questionnaire were invited to complete a self-administered questionnaire online once they returned home. Those who did not have internet or computer access were mailed a paper questionnaire. The on-site survey instrument consisted primarily of demographic and weather questions and was used as a means to interact with a random sample of travelers across Maine and increase the likelihood that visitors would complete the longer self-administered questionnaire (Dillman, Smyth, & Christian, 2014). Demographic data and weather perceptions were collected on both the front-end and self-administered survey instruments to assess non-response bias. The self-administered questionnaire consisted of five sections: (1) basic trip information, (2) the impacts of weather on their current trip and potential future trips, (3) accommodations and spending, (4) activities and climate change beliefs, and (5) demographic information. Questions on the survey were created after reviewing the relevant literature and using the question wording of studies that were measuring similar concepts and values (Brownlee et al., 2013; De Urioste-Stone et al., 2016; Stynes & White, 2006).

Data were collected between May–November 2015 throughout the State of Maine. A two-stage cluster probability sampling design was utilized (Scheaffer, Mendenhall, Ott, & Gerow, 2012), with the first phase consisting of a simple random sample of locations-times (cluster), and the second phase being the random sample visitors chosen from each cluster. First, we selected popular tourist locations throughout Maine to administer surveys, then random dates to survey were chosen, and finally random visitor groups were selected once on-site using systematic probability sampling. Trained survey administrators approached random groups walking by, and interviewed one person from each group (whoever had the most recent birthday) if willing. At the end of the survey instrument, tourists were asked for contact information in either the form of a mailing address or email address in order to send reminders about the self-administered survey instrument, as suggested by the Tailored Survey Design Method (Dillman et al., 2014). All onsite data were collected electronically on iPads using SurveyGizmo software.

After the onsite interview was completed, the visitor was given a postcard with a link to an online self-administered questionnaire and a personalized access code. All tourists who provided email or mailing addresses received up to three follow-up reminders. Reminders were sent every other week. Of 1712 onsite survey instruments completed by tourists, 704 respondents completed the self-administered questionnaire (41.1% response rate). A total of 688 responses were online (97.7%), and 16 responses (2.3%) were returned on hard copies. Since this study has defined that a tourist is someone who stays away from their permanent residence for at least one night, day visitors were not included in this study.

Data analysis

Using SPSS 22.0, a multivariate two-step cluster analysis was run to segment Maine visitors. Cluster analysis categorizes individuals into similar clusters based on sociodemographic characteristics, psychological factors, or behavior, so that those within a group are more similar to each other than those in other groups (Wedel & Kamakura, 2000). This analysis has been widely used in a variety of tourism and recreation contexts to understand group differences in tourist behavior (e.g. Rastogi, Harikrishna, & Patil, 2015), travel motivations (e.g. Bicikova, 2014; Chiang, Wang, Lee, & Chen, 2015), and management opinions (e.g. Hall, Seekamp, & Cole, 2010).

Clustering was based off of the number of nature-based activities visitors participated in or planned to participate in, and whether their primary recreational activity was nature-based or not. These two variables were created from a question that asked visitors: 'Which recreation activities did you participate in during this trip?' Visitors were asked to circle either 'participated,' 'planned to but could not,' or 'not interested on this trip' for each activity. Furthermore, respondents were asked to check which activity was their primary activity. Activities were categorized into whether they were completely nature-based, somewhat nature-based, or non-nature-based (Table 1). Activities were added to determine the number of nature-based activities visitors participated in or planned to participate in but could not.

This analysis yielded three segments of Maine visitors: nature-based specialists, nature-based generalists, and non-nature-based tourists. A recent study examining tourists in Norway used the terms 'specialist' and 'generalist' to segment nature-based tourists based

Table 1. Categorization of types of activities listed.

Categories	Activities
Not nature-based (0)	Arts or cultural activity, concert or festival, nightlife, shopping
Somewhat nature-based (0.5)	Golfing, picnicking, sightseeing/driving for pleasure
Nature-based (1)	Backpacking/hiking, biking, bird watching, boating, camping, canoeing/kayaking, climbing, fishing, hunting, viewing wildlife

Table 2. Categorization of questions on climate change perceptions and willingness to engage.

Category	Statements	α
General climate change belief/concern	I believe climate change is happening	0.923
	I am concerned about climate change	
Concern over the impact of climate change to tourism in Maine	I am concerned about the impacts of climate change to tourism in Maine	0.599
	The recreational activities that I enjoy in Maine would be at risk if local climate conditions were to change	
Civic engagement: education outreach	I am interested in learning more about the impacts of local climate change in Maine	0.817
	I would be willing to educate others about local climate change	
Civic engagement: donating money	I would be willing to donate money to reduce my carbon footprint when traveling to Maine	0.958
	I would be willing to donate money to help deal with the impacts from climate change in Maine	

on motivations. Specialists were defined as those who reported nature as the most important influence on their travel, whereas nature was not the most important factor for generalists (Mehmetoglu, 2005). This study instead segmented based on activities, and thus defined specialists as those who participated in fewer outdoor activities, and generalists as those who participated in more outdoor activities. These terms were used because the definition of a specialist is someone who is dedicated to one pursuit, while generalists are defined as those who have knowledge, skills, or interests in a variety of fields.

To test for differences between segments for weather sensitivities, the 5-point Likert scale was collapsed into three categories: not influential, slightly/moderately influential, and very/extremely influential. Chi-squares were run with Cramer's V for effect size. Adjusted standardized residuals (ASR) were used as a post-hoc, with those two standard deviations or more away from the expected mean noted.

To test for differences between the segments for climate change perceptions, eight questions on climate change perceptions were collapsed into four categories, with each category being the mean of two items (Table 2). Levene's statistic was used to first test the assumption of equal variances of groups. ANOVAs were used to compare for significant differences between the segments, and Tukey's Post Hoc were used if variances were equal, and Welch's ANOVA with Games-Howell Post Hoc were utilized for those with unequal variances (Vaske, 2008). Additionally, eta-squares (η^2) were used to examine the effect size.

Non-response bias was assessed by using Pearson's chi-square test of independence (x^2) to compare those who responded to the intercept survey ($n = 1712$) with those who responded to the self-administered survey ($n = 704$). These comparisons yielded no significant differences between those who completed the onsite and the self-administered survey instruments for age ($x^2 = 4297.16$, 4347 df, $p = .70$), gender ($x^2 = .02$, 1 df, $p = .89$), education ($x^2 = 40.52$, 42 df, $p = .54$), number of people in the travel party ($x^2 = 211.47$, 288 df, $p = 1.00$), number of nights ($x^2 = 653.15,675$ df, $p = .72$), and the importance of the expected weather ($x^2 = 12.14$, 16 df, $p = .73$).

Results

Segmentation

Visitors were segmented into three groups based on their activities participated in, with a silhouette of 0.6. As outlined in Table 3, the first group was labeled 'non-nature-based tourists,'

Table 3. Descriptions and characteristics of the three clusters. The first input is on a scale from 0 to 2, 0 meaning not nature-based, and 2 representing entirely nature-based.

Cluster	1	2	3
Label	Non-nature-based tourists	Nature-based generalists	Nature-based specialists
Description	Tourists who tended to participate in fewer nature-based activities, and have primary recreational activities that were not nature-based	Tourists who participated in the most nature-based activities, and the majority's primary activity was nature based	Tourists who did not participate in as many nature-based activities as the nature-based generalists, but all had a nature-based primary activity
Size	50.6% (356)	16.2% (114)	33.2% (234)
Inputs	Primary activity nature-based: 0.73	Primary activity nature-based: 1.89	Primary activity nature-based: 2.00
	Mean number of nature-based activities: 2.71	Mean number of nature-based activities: 7.01	Mean number of nature-based activities: 3.19

who tended to participate in fewer outdoor activities (mean = 2.71) and have their primary activity not nature-based. 'Nature-based generalists' were individuals who participated in many nature-based activities (mean = 7.01) and the majority had a nature-based activity as their primary activity. Finally, 'nature-based specialists' are those who participated in fewer outdoor activities (mean = 3.19), but everyone had a primary activity that was entirely nature-based. Slightly over half of Maine tourists were non-nature-based (50.6%), while the nature-based tourists were more specialists (33.2%) than generalists (16.2%).

Visitor profile

Tourists in Maine surveyed were predominately female (58.8%) and well educated (74.3% having a bachelor's degree or higher) (see Table 4); this fits with the general profile of Maine tourists found by the Maine Office of Tourism (Maine Office of Tourism, 2017, p. 109). Additionally, 46.0% of visitors were from the northeast region (See Figure 1), and 11.5% were Maine residents.

ANOVA elicited no statistically significant differences between the segments for education, and chi-square elicited no significant difference in gender at $\alpha = 0.05$. Region was significantly different between non-nature-based tourists and nature-based specialists, with the specialists having more Maine residents. The length of stay was different between non-nature-based tourists and nature-based generalists and specialists, with the non-nature-based tourists tending to have shorter trips. Additionally, nature-based generalists and specialists tended to have higher rates of overnight accommodation in RV and tent campgrounds, while non-nature-based tourists had higher rates of staying in hotels/motels/lodges. Age is also significantly different between non-nature-based tourists and generalists and specialists ($p < 0.001$), with the non-nature-based tourists being slightly older.

Weather sensitivity

When asked to rate the influence of five different weather conditions on their current trip, nature-based generalists perceived all weather conditions to be the most influential (Table 5). Nature-based specialists and non-nature-based tourists had similar responses, and were only significantly different for the influence of precipitation, with specialists reporting that it is more influential than non-nature-based tourists. Results show a relatively low influence of weather across all segments. However, when asked about the

Table 4. A profile of tourists who responded to the self-administered survey, broken down by group. Numbers for the segments and sample average are expressed as percentages.

Demographic and trip characteristics	Sample average (n = 704)	Non-nature-based (n = 356)	Generalists (n = 114)	Specialists (n = 234)	Chi-square	ANOVA F	Sig
Gender					0.15		.93
Male	41.2	40.8	40.4	42.2			
Female	58.8	59.2	59.6	57.8			
Age in years						13.97	<.01
Mean	54.2	56.9	50.9	51.6			
Travel group size						7.84	<.01
Mean	2.75	2.58	3.37	2.68			
Education					0.29		.75
High school or less	6.9	8.1	4.5	6.2			
Some college	9.8	8.4	9.1	12.3			
2-year degree	8.9	8.1	14.5	7.5			
Bachelor's degree	33.1	32.5	28.2	36.6			
Graduate degree	41.2	42.9	43.7	37.4			
Overnight lodging							
Hotel/motel	65.4	76.7	50.9	55.3	72.70		<.01
Friends/seasonal	20.0	18.7	17.3	23.2			
RV camping	6.7	3.2	11.8	9.6			
Tent camping	7.9	1.4	20.0	11.8			
Length of stay					12.05		<.01
1-3 Nights	41.9	50.6	19.3	39.7			
4-7 Nights	43.9	39.3	64.9	40.6			
8-14 Nights	10.8	7.0	10.5	16.7			
15+ Nights	3.4	3.1	5.3	3.0			
Region from					7.17		<.01
Maine	11.5	7.5	9.3	18.8			
Northeast	46.0	44.2	50.9	46.4			
South	21.4	23.7	19.4	18.8			
Midwest	9.6	10.7	9.3	8.0			
West	5.5	6.6	5.6	3.6			
International	6.0	7.2	5.6	4.5			

importance of the weather on their trip in general, only 13.7% of tourists believed weather was unimportant or very unimportant. Additionally, 22.2% of non-nature-based tourists, 35.1% of nature-based generalists, and 28.2% of nature-based specialists changed their travel or recreation plans on their trip due to weather.

When looking at tourist by lodging type rather than recreational activities, tent campers were significantly more likely to say the overall weather was very or extremely influential in choosing their destination and during their travels (Table 6). Those who were staying at the residence of friends/family or their own seasonal residence thought the overall weather during their trip and the expected weather were less important. Additionally, tent campers thought precipitation was very/extremely influential at a higher rate than those in other accommodations, but there were no significant differences in the influence of temperature, sunshine, or wind by overnight accommodations.

Climate change perceptions

Most of the tourists to Maine tended to believe in climate change and be concerned about it (mean of 3.87, on a scale from 1-5, with 1 being strongly disagree and 5 being strongly agree). Additionally, many tourists were concerned about climate change in

Table 5. Perceived influence of weather on Maine tourists' trips by cluster.

Weather variable		Avg (%)	Non-nature (%)	Gene-ralists (%)	Speci-alists (%)	Chi-square	Sig	Cramer's V
Precipitation	Not influential	43.2	49.3*	29.8*	40.6	17.962	.001	.113
	Slightly/moderately	40.0	35.8*	54.4*	39.3			
	Very/extremely	16.8	14.9	15.8	20.1			
Sunshine	Not influential	37.3	40.6	26.3*	37.8	10.282	.036	.086
	Slightly/moderately	40.8	36.4*	49.1*	43.3			
	Very/extremely	21.9	23.0	24.6	18.9			
Max temperature	Not influential	39.7	41.6	26.8*	42.9	9.630	.047	.083
	Slightly/moderately	42.5	41.3	50.0	40.8			
	Very/extremely	17.8	17.1	23.2	16.3			
Min. temperature	Not influential	47.4	48.6	36.8*	50.9	7.568	.109	
	Slightly/moderately	42.1	42.3	50.0	37.9			
	Very/extremely	10.5	9.1	13.2	11.2			
Wind speed	Not influential	58.7	60.5	50.4	59.9	8.943	.063	
	Slightly/moderately	35.7	34.7	46.0*	32.3			
	Very/extremely	5.6	4.8	3.5	7.8			
Importance of the actual weather (general)	Unimportant	13.8	13.8	11.4	15.0	9.741	.045	.118
	Neither important nor unimportant	28.3	32.7*	20.2*	25.8			
	Important	57.8	53.5*	68.4*	59.2			
Importance of the expected weather in choosing destination	Unimportant	24.5	23.0	22.1	28.0	8.980	.062	
	Neither important nor unimportant	34.7	38.2*	26.5*	33.2			
	Important	40.8	38.8	51.3*	38.8			

Note: Degrees of freedom = 4 for all chi-square tests.
*Indicates adjusted standardized residual (ASR) >1.96.

Maine as well (mean of 3.36). Overall, nature-based specialists and nature-based generalists had the highest belief and concern about climate change in general and in Maine, while non-nature-based tourists had lower belief and concern (Table 7).

Additionally, some tourists indicated they would be willing to participate in education outreach (mean of 2.85, on a scale from 1 to 5, with 1 being strongly disagree and 5 being strongly agree) or donate money (mean of 2.73) to help local climate change. Although levels of belief and concern for climate change were not statistically different between nature-based generalists and specialists, generalists had statistically higher willingness to engage in climate change civic action for both donating money and education outreach.

Discussion and conclusions

The goal of this study was to understand differences in weather sensitivities and climate change perceptions across different types of tourists to the same area. Climate change is already impacting tourism globally, and is predicted to continue altering where and when visitors travel (Gössling et al., 2012). Since tourism is a significant contributor to many economies, potential future changes to tourism are important to understand. However, tourists are not a homogenous group and could respond differently to changes in weather and climate.

Across all five weather variables, generalists were the most impacted by the weather, and non-nature-based tourists the least. Although nature-based generalists perceive the weather to be the most influential, they were also adapting the most by changing the

Table 6. Perceived influence of weather on Maine tourists' trips by type of overnight accommodations.

Weather variable		Hotel, motel (%)	Friends/ seasonal (%)	RV camp (%)	Tent camp (%)	Chi-Square	Sig	Cramer's V
Precipitation	Not influential	44.2	48.2	41.3	29.6*	17.009	.009	.111
	Slightly/moderately	40.8	38.7	43.5	35.2			
	Very/extremely	15.0	13.1	15.2	35.2*			
Sunshine	Not influential	37.3	39.7	41.3	33.3	5.209	.517	
	Slightly/moderately	41.3	39.7	41.3	33.3			
	Very/extremely	21.3	20.6	17.4	33.3*			
Max temperature	Not influential	38.3	50.4*	39.1	28.8	10.309	.112	
	Slightly/moderately	43.8	32.3*	45.7	50.0			
	Very/extremely	17.9	17.3	15.2	21.2			
Min. temperature	Not influential	49.2	50.4	43.5	35.2*	7.024	.319	
	Slightly/moderately	41.6	36.3	45.7	50.0			
	Very/extremely	9.2	13.3	10.9	14.8			
Wind speed	Not influential	58.5	63.7	56.8	55.6	8.728	.189	
	Slightly/moderately	37.0	27.4*	40.9	37.0			
	Very/extremely	4.5	8.9*	2.3	7.4			
Importance of the actual weather (general)	Unimportant	11.9*	20.6*	17.4	13.0	16.549	.011	.110
	Neither important nor unimportant	29.5	31.6	23.9	13.0*			
	Important	58.6	47.8*	58.7	74.1*			
Importance of the expected weather in choosing destination	Unimportant	21.5*	36.5*	23.9	22.2	23.073	.001	.130
	Neither important nor unimportant	34.5	38.0	37.0	24.1*			
	Important	43.9*	25.5*	39.1	53.7*			

Note: Degrees of freedom = 6 for all chi-square tests.
*Indicates adjusted standardized residual (ASR) >1.96.

activities they participated in. Overall, when asked about individual weather factors, tourists reported a small influence across all variables, with sunshine being the most influential across all groups, and wind being the least influential. The findings that wind is the least important weather variable fits with previous studies, although depending on the location, others have found rain or temperature to be more influential than sunshine (Hewer et al., 2015; Scott et al., 2008; Steiger et al., 2016).

However, when asked about the importance of the weather on their trip in general, only 13.8% believed weather was unimportant. So although tourists may perceive the weather to be influential as a whole, individual components of the weather may not be seen as particularly important. Almost a third of tourists to Maine changed their plans due

Table 7. Climate change concern and willingness to engage in civic action. Items represent a mean of two questions measuring the same concept, on a scale from 1 (strongly disagree) to 5 (strongly agree).

	Avg.	Non- nature	Gene- ralists	Speci- alists	Levene Stat (sig)	ANOVA F/Welch's F (sig)	Effect size
General climate change belief/ concern	3.87	3.73[a]	4.00[a,b]	4.02[b]	6.97 (.001)	Welch: 6.05 (.003)	ω^2 .015
Concern over the impact of climate change to tourism in Maine	3.36	3.24[a]	3.51[b]	3.46[b]	0.48 (.619)	F: 6.38 (.003)	η^2.019
Civic engagement: education outreach	2.85	2.70[a]	3.24[b]	2.88[a]	0.67 (.514)	F: 12.92 (<.001)	η^2.038
Civic engagement: donating money	2.73	2.56[a]	3.09[b]	2.80[c]	1.15 (.318)	F: 12.42 (<.001)	η^2.036

[abc]Means followed by different letters are statistically significant at $\alpha = 0.05$ found using Tukey's Post Hoc test for equal variances, and Games-Howell when variances were unequal.

to the weather, and a previous study in Norway also found that a third of visitors changed plans because of the weather (Denstadli et al., 2011). While nature-based generalists reported the largest influence of weather, they also may be the most adaptable, since they participated in more activities and changed the activities they participated in the most. This is important because there are micro-climates within a single tourism destination, which provides tourists some capacity to adapt to weather; for instance, moving to the beach when temperatures are hotter than desired (Rutty & Scott, 2014). Although findings suggest that a changing climate could impact the visitation and satisfaction of nature-based generalists the most, they may be more adaptable than nature-based specialists. Thus, it may be beneficial to advertise a wide range of recreational activities available at the destination so tourists, particularly nature-based specialists, have an awareness of options to adapt to negative conditions.

People who were tent camping for their overnight accommodations were more likely to state the importance of the weather when selecting their destination, and also the importance of the actual weather during their trip. Thus, this group would be the most likely to change future behavior due to the weather. However, the only weather variable that had more influence on campers was precipitation. With increased precipitation expected in Maine due to climate change, particularly during popular camping seasons, fewer campers may choose to visit the area. Campgrounds could adapt to this weather sensitivity by adding more awnings, sheltered camp spots, or sheltered picnic areas so visitors who prefer camping are less influenced by increased precipitation. Those who were staying with friends/family thought the weather was the least important when selecting their destination and during the actual trip. This analysis also showed the difference between tent and RV campers, as RV campers reported lower levels of influence of weather than tent campers. However, Hewer et al. (2015) found that the weather's impact on campers even varied based on location, with beach campers tending to be more sensitive to weather than forest campers. This study did not ask people where they were camping, and based on the landscape of the state it is likely a mix of forest and coastal campers.

These findings are relevant for tourism destinations as well as those interested in climate change mitigation. Across all three groups, tourists tend to be concerned about the impacts to tourism and recreational activities in Maine. Climate change perceptions and beliefs among different groups can be important to understand in trying to influence behavior (Brownlee, Hallo, Wright, Moore, & Powell, 2013; Brownlee, Powell, & Hallo, 2013). Additionally, some tourists indicated they would be willing to donate money or participate in educational efforts to help local climate change. The mean responses of 2.73 (donating money) and 2.85 (education) by Maine tourists were very similar to results found when surveying visitors to a botanical garden in South Carolina, USA (Brownlee et al., 2013). Although specialists have slightly higher belief/concern in climate change, generalists have the highest stated willingness to donate money or educate others about climate change. Since nature-based generalists already report the highest level of willingness to donate money and educate, these tourists would be the best group to target for civic engagement on climate change. However, studies have shown that visitors' concern and awareness of climate change is positively correlated with their willingness to participate in civic action, so long-term it may be beneficial to target climate change information at non-nature-based tourists, who show the lowest levels of climate change belief and concern (e.g. Brownlee et al., 2013).

This study does have limitations, namely that tourists were not given specific weather conditions to evaluate, so respondents may have interpreted weather questions differently. Additionally, tourists were not asked directly about future climate and weather scenarios. Rather, this study aimed to look at tourist groups and begin investigating differences in weather and climate change perceptions. Future research could embed choice experiments based on climate change scenarios (e.g. Pröbstl-Haider & Haider, 2013; Pröbstl-Haider, Haider, Wirth, & Beardmore, 2015) and segment based on intended behavior changes to better understand the characteristics of tourists who will be most impacted by climate change. Segmenting visitors on intended behavior, values, and motivations could provide further insight into different tourist types and how behavior might differ with future climate change. Segmenting tourists was advantageous to explore differences in weather sensitivities and climate change perceptions and could be useful to further investigate the effect of climate change on tourism.

Acknowledgements

The authors are thankful to the Maine Office of Tourism, Acadia National Park, Maine Bureau of Parks and Lands, Baxter State Park Authority, Bangor International Airport, and the Bethel Chamber of Commerce for their support and willingness to allow visitor surveys. The authors are also grateful to Lydia Horne, Matt Scaccia, and Ashley Cooper for their contribution to the data collection, to Washington State University for hosting the online survey, and to Dr Lena Le for reviewing the survey instrument. Special thanks to the anonymous reviewers for their helpful feedback.

Disclosure statement

No potential conflict of interest was reported by the authors.

Funding

This work was supported by the National Institute of Food and Agriculture, U.S. Department of Agricultre [grant number 1003857]; University of Maine Office of the President.

ORCID

Emily Wilkins (iD) http://orcid.org/0000-0003-3055-4808?
Sandra de Urioste-Stone (iD) http://orcid.org/0000-0002-7284-649X

References

Becken, S. (2012). Measuring the effect of weather on tourism: A destination- and activity-based analysis. *Journal of Travel Research, 52*(2), 156–167. doi:10.1177/0047287512461569

Becken, S., & Hay, J. E. (2007). *Tourism and climate change: Risks and opportunities.* Tonawanda, NY: Multilingual Matters.

Becken, S., & Wilson, J. (2013). The impacts of weather on tourist travel. *Tourism Geographies, 15*(4), 620–639. doi:10.1080/14616688.2012.762541

Bicikova, K. (2014). Understanding student travel behavior: A segmentation analysis of British university students. *Journal of Travel & Tourism Marketing, 31*(7), 854–867. doi:10.1080/10548408.2014.890154

Brownlee, M. T. J., Hallo, J. C., & Krohn, B. D. (2013). Botanical garden visitors' perceptions of local climate impacts: Awareness, concern, and behavioral responses. *Managing Leisure, 18*(2), 97–117. doi:10.1080/13606719.2013.752209

Brownlee, M. T. J., Hallo, J. C., Wright, B. A., Moore, D., & Powell, R. B. (2013). Visiting a climate-influenced national park: The stability of climate change perceptions. *Environmental Management, 52,* 1132–1148. doi:10.1007/s00267-013-0153-2

Brownlee, M. T. J., Powell, R. B., & Hallo, J. C. (2013). A review of the foundational processes that influence beliefs in climate change: Opportunities for environmental education research. *Environmental Education Research, 19*(1), 1–20. doi:10.1080/13504622.2012.683389

Cadarso, M.-Á., Gómez, N., López, L.-A., Tobarra, M.-Á., & Zafrilla, J.-E. (2015). Quantifying Spanish tourism's carbon footprint: The contributions of residents and visitors: a longitudinal study. *Journal of Sustainable Tourism, 23*(6), 922–946. doi:10.1080/09669582.2015.1008497

Chiang, C. C., Wang, M. Y., Lee, C. F., & Chen, Y. C. (2015). Assessing travel motivations of cultural tourists: A factor-cluster segmentation analysis. Journal of Information & Optimization Sciences, 36 (3), 269–282. doi:10.1080/02522667.2014.996028.

Choi, A. S., Ritchie, B. W., & Fielding, K. S. (2016). A mediation model of air travelers' voluntary climate action. *Journal of Travel Research, 55*(6), 709–723. doi:10.1177/0047287515581377

Dawson, J., & Scott, D. (2013). Managing for climate change in the alpine ski sector. *Tourism Management, 35,* 244–254. doi:10.1016/j.tourman.2012.07.009

De Urioste-Stone, S. M., Le, L., Scaccia, M. D., & Wilkins, E. (2016). Nature-based tourism and climate change risk: Visitors' perceptions in mount desert island, Maine. *Journal of Outdoor Recreation and Tourism.* Retrieved from https://doi.org/10.1016/j.jort.2016.01.003i

De Urioste-Stone, S. M., Scaccia, M. D., & Howe-Poteet, D. (2015). Exploring visitor perceptions of the influence of climate change on tourism at Acadia National Park, Maine. *Journal of Outdoor Recreation and Tourism, 11,* 34–43. doi:10.1016/j.jort.2015.07.001

Denstadli, J. M., Jacobsen, J. K. S., & Lohmann, M. (2011). Tourist perceptions of summer weather in Scandinavia. *Annals of Tourism Research, 38*(3), 920–940. doi:10.1016/j.annals.2011.01.005

Dillman, D. A., Smyth, J. D., & Christian, L. M. (2014). *Internet, phone, mail, and mixed-mode surveys: The tailored design method* (4th ed.). Hoboken, NJ: Wiley.

Explore Maine. (2014). Explore Maine by tourism region. Retrieved from http://www.exploremaine.org/region/index.shtml

Falk, M. (2014). Impact of weather conditions on tourism demand in the peak summer season over the last 50 years. *Tourism Management Perspectives, 9,* 24–35. doi:10.1016/j.tmp.2013.11.001

Fernandez, I. J., Schmitt, C. V., Birkel, S. D., Stancioff, E., Pershing, A. J., Kelley, J. T., ... Mayewski, P. A. (2015). *Maine's climate future: 2015 update.* Orono, ME: University of Maine.

Forests for Maine's Future. (2011). FAQ's about Maine's forests. Retrieved from http://www.forestsformainesfuture.org/forest-facts/

Gómez Martín, M. B. (2005). Weather, climate and tourism a geographical perspective. *Annals of Tourism Research, 32*(3), 571–591. doi:10.1016/j.annals.2004.08.004

Gössling, S., & Peeters, P. (2015). Assessing tourism's global environmental impact 1900-2050. *Journal of Sustainable Tourism, 23*(5), 639–659. doi:10.1080/09669582.2015.1008500

Gössling, S., Haglund, L., Kallgren, H., Revahl, M., & Hultman, J. (2009). Swedish air travellers and voluntary carbon offsets: Towards the co-creation of environmental value? *Current Issues in Tourism, 12*(1), 1–19. doi:10.1080/13683500802220687

Gössling, S., Scott, D., Hall, C. M., Ceron, J.-P., & Dubois, G. (2012). Consumer behaviour and demand response of tourists to climate change. *Annals of Tourism Research, 39*(1), 36–58. doi:10.1016/j.annals.2011.11.002

Hall, T. E., Seekamp, E., & Cole, D. (2010). Do recreation motivations and wilderness involvement relate to support for wilderness management? A segmentation analysis. *Leisure Sciences, 32*(2), 109–124. doi:10.1080/01490400903547096

Hewer, M. J., Scott, D., & Gough, W. A. (2015). Tourism climatology for camping: A case study of two Ontario parks (Canada). *Theoretical Applied Climatology, 121*(3–4), 401–411. doi:10.1007/s00704-014-1228-6

Intergovernmental Panel on Climate Change. (2013). Summary for policymakers. In T. F. Stocker, D. Qin, G.-K. Plattner, M. Tignor, S. K. Allen, J. Boschung, A. Nauels, Y. Xia, V. Bex, & P. M. Midgley (Eds.), *Climate Change 2013: The Physical Science Basis*. Contribution of Working Group I to the Fifth Assessment Report of the Intergovernmental Panel on Climate Change (pp. 2–6). Cambridge and New York, NY: Cambridge University Press.

Jeuring, J. H. G., & Peters, K. B. M. (2013). The influence of the weather on tourist experiences: Analysing travel blog narratives. *Journal of Vacation Marketing, 19*(3), 209–219. doi:10.1177/1356766712457104

Leiper, N. (1979). The framework of tourism: Towards a definition of tourism, tourist, and the tourist industry. *Annals of Tourism Research, 6*(4), 390–407.

Loomis, J. B., & Richardson, R. B. (2006). An external validity test of intended behavior: Comparing revealed preference and intended visitation in response to climate change. *Journal of Environmental Planning and Management, 49*(4), 621–630. doi:10.1080/09640560600747562

Maine Office of Tourism. (2016). 2015 Maine tourism highlights. Retrieved from https://visitmaine.com/assets/downloads/2015_FactSheet.pdf

Maine Office of Tourism. (2017). Maine office of tourism visitor tracking research: 2016 Calendar Year Annual Report. Retrieved from https://visitmaine.com/assets/downloads/2016-MOT-Annual-Report.pdf

McKercher, B., Prideaux, B., Cheung, C., & Law, R. (2010). Achieving voluntary reductions in the carbon footprint of tourism and climate change. *Journal of Sustainable Tourism, 18*(3), 297–317. 10.1080/09669580903395022

McKercher, B., Shoval, N., Park, E., & Kahani, A. (2015). The [limited] impact of weather on tourist behavior in an urban destination. *Journal of Travel Research, 54*(4), 442–455. doi:10.1177/0047287514522880

Mehmetoglu, M. (2005). A case study of nature-based tourists: Specialists versus generalists. *Journal of Vacation Marketing, 11*(4), 357–369. doi:10.1177/1356766705056634

Mehmetoglu, M. (2007). Nature-based tourists: The relationship between their trip expenditures and activities. *Journal of Sustainable Tourism, 15*(2), 200–215. doi:10.2167/jost642.0

Mieczkowski, Z. (1985). The tourism climate index: A method of evaluating world climates for tourism. *The Canadian Geographer, 29*(3), 220–233.

Moreno, A., & Amelung, B. (2009). Climate change and tourist comfort on Europe's beaches in summer: A reassessment. *Coastal Management, 37*, 550–568.

Moreno, A., Amelung, B., & Santamarta, L. (2009). Linking beach recreation to weather conditions: A case study in Zandvoort, Netherlands. *Tourism in Marine Environments, 5*(2–3), 111–119.

Morgan, R., Gatell, E., Junyent, R., Micallef, A., Ozhan, E., & Williams, A. T. (2000). An improved user-based beach climate index. *Journal of Coastal Conservation, 6*, 41–50.

Pröbstl-Haider, U., & Haider, W. (2013). Tools for measuring the intention for adapting to climate change by winter tourists: Some thoughts on consumer behavior research and an empirical example. *Tourism Review, 68*(2), 44–55. Retrieved from https://doi.org/10.1108/TR-04-2013-0015

Pröbstl-Haider, U., Haider, W., Wirth, V., & Beardmore, B. (2015). Will climate change increase the attractiveness of summer destinations in the European Alps? A survey of German tourists. *Journal of Outdoor Recreation and Tourism, 11*, 44–57. Retrieved from https://doi.org/10.1016/j.jort.2015.07.003

Rastogi, R., Harikrishna, M., & Patil, A. (2015). Segmentation analysis of domestic tourists – A case study. *KSCE Journal of Civil Engineering, 19*(5), 1509–1522. doi:10.1007/s12205-015-0673-9

Roser-Renouf, C., Maibach, E. W., Leiserowitz, A., & Zhao, X. (2014). The genesis of climate change activism: From key beliefs to political action. *Climatic Change, 125*, 163–178. doi:10.1007/s10584-014-1173-5

Rutty, M., & Scott, D. (2010). Will the Mediterranean become "too hot" for tourism? A reassessment. *Tourism and Hospitality Planning and Development, 7*(3), 267–281. doi:10.1080/1479053X.2010.502386

Rutty, M., & Scott, D. (2014). Thermal range of coastal tourism resort microclimates. *Tourism Geographies, 16*(3), 346–363. doi:10.1080/14616688.2014.932833

Rutty, M., & Scott, D. (2016). Comparison of climate preferences for domestic and international beach holidays: A case study of Canadian travelers. *Atmosphere, 7*(30), 1–12. doi:10.3390/atmos7020030

Scheaffer, R. L., Mendenhall, W., III, Ott, R. L., & Gerow, K. G. (2012). *Elementary survey sampling* (7th ed.). Boston, MA: Cengage Learning.

Scott, D., Dawson, J., & Jones, B. (2008). Climate change vulnerability of the US northeast winter recreation - tourism sector. *Mitigation and Adaptation Strategies for Global Change, 13*(5–6), 577–596. doi:10.1007/s11027-007-9136-z

Scott, D., Gössling, S., & Hall, C. M. (2012). International tourism and climate change. *WIREs Climate Change, 3*, 213–232. doi:10.1002/wcc.165

Scott, D., Gössling, S., & de Freitas, C. (2008). Preferred climates for tourism: Case studies from Canada, New Zealand and Sweden. *Climate Research, 45*(December), 61–73. doi:10.3354/cr00774

Scott, D., Jones, B., & Konopek, J. (2007). Implications of climate and environmental change for nature-based tourism in the Canadian Rocky Mountains: A case study of Waterton Lakes National Park. *Tourism Management, 28*(2), 570–579. doi:10.1016/j.tourman.2006.04.020

Scott, D., McBoyle, G., & Schwartzentruber, M. (2004). Climate change and the distribution of climatic resources for tourism in North America. *Climate Research, 27*, 105–117.

Segerstedt, A., & Grote, U. (2016). Increasing adoption of voluntary carbon offsets among tourists. *Journal of Sustainable Tourism, 24*(11), 1541–1554. doi:10.1080/09669582.2015.1125357

Sharp, H., Grundius, J., & Heinonen, J. (2016). Carbon footprint of inbound tourism to Iceland: A consumption-based life-cycle assessment including direct and indirect emissions. *Sustainability, 8* (1147). doi:10.3390/su8111147

Smith, K. (1993). The influence of weather and climate on recreation and tourism. *Weather, 48*, 398–404.

Steiger, R., Abegg, B., & Jänicke, L. (2016). Rain, rain, go away, come again another day. Weather preferences of summer tourists in mountain environments. *Atmosphere, 7*(63), 1–12. doi:10.3390/atmos7050063

Stynes, D. J., & White, E. M. (2006). Reflections on measuring recreation and travel spending. *Journal of Travel Research, 45*, 8–16. doi:10.1177/0047287506288873

Vaske, J. J. (2008). *Survey research and analysis: Applications in parks, recreation and human dimensions*. State College, PA: Venture Publishing, Inc.

Wedel, M., & Kamakura, W. A. (2000). *Market segmentation: Conceptual and methodological foundations*. New York, NY: Springer.

Wight, P. A. (2001). Ecotourists: Not a homogeneous market segment. In D. B. Weaver (Ed.), *The encyclopedia of ecotourism* (pp. 37–62). New York, NY: CABI Publishing.

World Travel & Tourism Council. (2017). Travel & tourism: Economic impact 2017 United States. Retrieved from https://www.wttc.org/-/media/files/reports/economic-impact-research/countries-2017/unitedstates2017.pdf

Micro-level assessment of regional and local disaster impacts in tourist destinations

Jürgen Schmude, Sahar Zavareh, Katrin Magdalena Schwaiger and Marion Karl

ABSTRACT

The tourism sector faces severe challenges due to the economic impacts from changing natural environments as seen with the increased frequency of natural disasters. Therefore, analyses of disaster impacts models are necessary for managing successful tourism recovery. Typically, disaster assessments are conducted on a countrywide level, which can lead to imbalanced recovery processes, and a distorted distribution of recovery financing or subsidies. We address the challenges of recovery using the tourism disaster management framework by Faulkner. To calculate precise damage assessments, we develop a micro-level assessment model to analyze and understand disaster impacts at the micro-level supporting tourism recovery in an affected destination. We examine economic consequences of a disaster at a small regional scale arguing recovery from a natural disaster is more difficult in individual areas because of differences in geographic location or infrastructure development. The island of Dominica is chosen as an example for the model using statistical data from the tourism sector to outline and detail the consequences of a disaster specifically for communities. The results highlight the importance of damage assessments on a small-scale level, such as communities in order to distinguish between individual regions facing severe changes for resident livelihoods and the local tourism sector. We argue that only after identifying regional impacts it is possible to apply adequate governmental subsidies and development strategies for a country's tourism sector and residents in a continuously changing environment in the hopes of mitigating future financial losses and future climate change impacts.

摘要

随着自然灾害频繁发生, 自然环境的变化带来的经济影响, 旅游部门面临着严峻的挑战。因此, 对灾害影响模型的分析是成功复苏旅游业的必要条件。通常情况下, 灾难评估是在全国范围内进行的, 这可能导致复苏过程的不平衡, 以及资金或补贴分配的扭曲。我们利用福克纳(2001)的旅游灾害管理框架来应对旅游复苏的挑战。为了计算精确的损失估价, 我们开发了一个微观层面的评估模型, 分析和理解受灾目的地微观层面上支持旅游业复苏的灾难影响。我们在一个小规模的区域范围内研究灾难的经济后果, 认为由于地理位置或基础设施发展的不同, 个别地区自然灾害的恢复更加困难。多米尼加岛被选为该模型的范例, 利用旅游部门的统计数据, 对一场专门针对社区的灾难后果进行描绘和详细说明。这些结果突出了对小规模（例如社区）的损害评估的重要性，以便区分居民生计面临严重变化的地区和当地旅游业面临严重变化的地区。我们认为，只有在确定了区域影响之后，才有可能在不断变化的环境中为一个国家的旅游部门和居民提供适当的政府补贴和发展战略，以减轻未来的财政损失和气候变化的影响。

Introduction

Natural disasters range from volcanism, earthquakes, hurricanes and tropical storms bringing destruction and severe consequences such as flooding, landslides, or built infrastructure destroyed to the impacted region. The severity of disaster impacts largely depends on a country's wealth, the state of its economic development and diversification (Hallegatte & Ghil, 2008; Heger, Julca, & Paddison, 2008; Kahn, 2005; Neumayer, Plümper, & Barthel, 2014; Toya & Skidmore, 2005). Impacts of natural disasters can be assessed as four types of damages: direct damages referring to impacts on housing or roads; indirect damages such as job losses or health deterioration; quantitative damage (tangible effects) related to the absolute monetary losses caused by destruction of buildings; and qualitative damage (intangible effects) signifying insecurity or social disruption (ECLAC, 2003; Hallegatte & Przyluski, 2010; Merz, Kreibich, Thieken, & Schmidtke, 2004). Countries with greater economic development and diversity at the time of a disaster tend to experience lower disaster losses, whereas countries largely dependent on tourism alone are more likely to be impacted by disaster events (Kim & Marcouiller, 2015; Noy, 2009). Moreover, natural disasters are likely to increase due to the challenges brought about by climate change resulting from changes in air temperatures, precipitation rates, sea level rise, an increased frequency of heat waves, hurricanes, tropical storms, and higher tropical-cyclone-related rainfall rates (IPCC, 2014; Knutson et al., 2010; Trenberth, 2011).

Small island developing states ('SIDS') are especially vulnerable to hurricane and tropical storm hazards (Briguglio, 2003; Collymore, 2011; Forbes, James, Sutherland, & Nichols, 2013; IPCC, 2007; Méheux, Dominey-Howes, & Lloyd, 2007; Mimura & Nurse, 2017). The likelihood of natural disasters in small islands is attributed to their limited resources, size, their location surrounded by ocean waters, and rising sea levels (Ferdinand, Haynes, & Richards, 2014; Mimura et al., 2007; Pelling & Uitto, 2001). SIDS are some of the most disadvantaged places affected by global warming (IPCC, 2007), and viewed as one of Earth's barometers of climate change (Kelman & West, 2009). Furthermore, climate change is an important factor leading to increased hazard exposure in the tourism sector for island destinations (Becken, Mahon, Rennie, & Shakeela, 2014; Tsao & Ni, 2016). Many SIDS are dependent on tourism and are susceptible to the consequences of climate change due to increased damages to tourism infrastructures (Becken & Hay, 2007; Ibarrarán, Ruth, Ahmad, & London, 2009). This is referred to as 'endangered future' for tourist destinations because tourists are less likely to choose destinations where they perceive increased hurricane risk (Forster, Schuhmann, Lake, Watkinson, & Gill, 2012). This further illustrates the high influence disasters have on the tourism industry and the reputation of a destination (Durocher, 1994; Méheux & Parker, 2006).

Most research on disaster impact assessment is from a macroeconomic perspective only addressing consequences and impact analysis, yet tourism-based communities are rarely discussed independently in dealing with natural disasters (Kim & Marcouiller, 2015;

Tsai & Chen, 2011). Studies largely focus on entire countries without providing more detail on specific damages for smaller regional scales while omitting regional particularities of a countrywide disaster. Misleading conclusions about the gravity of the disaster are likely if damages are solely assessed at a national level or do not consider the scale of impacts. These kinds of analysis do not provide the disparity of damages created from the disaster where some areas are more affected than others, thus making it more difficult to distribute governmental and financial assistance, or other available subsidies to residents. Financial assistance must not only be given to the entire country, but should be distributed accordingly to those areas with the worst impact in a time-efficient manner to avoid additional disaster aid barriers created by long negotiations with donors (Pelling, Özerdem, & Barakat, 2002). While dealing with the aftermath of disasters requires structured and accurate governance, sub-national distribution of economic support is necessary (Scolobig et al., 2014; Strobl, 2012).

To date, developing models estimating damages at local scales has not taken place within tourism research. We address this challenge by developing a micro-level assessment model (MLAM) to determine precise damage assessments. Dominica was chosen as study area due to the significance of the tourism sector for the island, the increased likelihood of hurricanes and tropical storms, as well as the additional dangers faced from climate change (e.g. sea level rise). Pre-disaster adaptation measures generally are preferable, such as improved hurricane warnings may decrease damages from disasters (Sadowski & Sutter, 2005), or the use of pre-disaster risk estimations to manage post-disaster losses (Tsai & Chen, 2011; Tsao & Ni, 2016). Implementing such measures is costly and requires the consideration of more than one hazard exposure (Anderson, 1995). We highlight the importance of regionally differentiated impact analysis using possible forced adaptation measures as described by Tervo-Kankare, Kajan, and Saarinen (2016), because they can be viewed as 'benefits' for future disasters. While MLAM is a tool that primarily aids in post-disaster damage evaluation, it may be relevant for governments and institutions supporting appropriate disaster recovery processes for communities to decrease future vulnerabilities. The model is described in greater detail first by providing a literature review of disaster impact assessments, second a discussion of the study area and disaster implications, and third by identifying the methodology used to determine the financial consequences for the tourism sector to draw conclusions from the disaster and recovery process.

Literature review

Islands are known for their vulnerabilities to natural hazards creating substantial risks for tourism industry economies (Becken et al., 2014). Although there are myriad forms of natural disaster impact assessments in the literature, the tourism sector has often not been a focal point of research in the context of natural disasters (Kim & Marcouiller, 2015; Tsai & Chen, 2011). Current analysis is conducted from a macroeconomic perspective estimating general impacts on local economies in terms of costs and consequences of individual disasters on a wide range of scales (Baade, Baumann, & Matheson, 2007; Horwich, 2000; Selcuk & Yeldan, 2001; Vigdor, 2008), including entire countries or even continents (Cavallo, Powell, & Becerra, 2010; Jovel, 1989). The most common variables used for economic damage assessments are the number of people killed or number of persons affected (Jovel,

1989, Kim & Marcouiller; 2015; Noy, 2009; Vigdor, 2008), the rise in external transfer payments and grants (West & Lenze, 1994), monetary damages (Baade, Baumann, & Matheson, 2007; Cavallo et al., 2010; Jovel, 1989; Kim & Marcouiller, 2015; West & Lenze, 1994), production losses (Hallegatte & Przylusky, 2010), macroeconomic growth rates or general changes in the GDP (Albala-Bertrand, 1993; Benson & Clay, 2004; Horwich, 2000; Strobl, 2012), and employment changes (Coffman & Noy, 2011; Ellson, Milliman, & Roberts, 1984; Ewing & Kruse, 2005).

There is a need for disaster phenomena research specifically related to tourism (Faulkner, 2001; Faulkner & Vikulov 2001; Ritchie, 2004). Lee and Chi (2013) provide examples as to how to analyze disaster impacts on tourism by measuring the change of annual visitors in Taiwan at scenic spots after the 1999 earthquake. Tsai, Wu, Wall, and Linliu (2016) conducted qualitative interviews with local communities in Taiwan measuring perceptions of tourism impacts on communities, thereby demonstrating that the impact of disasters on the tourism sector is critical to understanding tourism disaster recovery.

Apart from assessments of social components such as visitors or impacted community residents after a disaster, it is common in literature to use economic indicators for disaster impact analysis. Irrespective of a disaster impact, standard assessments of economic impacts (e.g. by the tourism industry) for a region are conducted with input–output (IO) or computable general equilibrium models (CGE) (Dwyer, Forsyth, & Spurr, 2004; Kumar & Hussain, 2014). Advantages and disadvantages of both approaches for disaster impact analysis have been assessed, and IO models in general are considered insufficient for estimating potential substitution effects for impacted regions (Koks et al., 2016). Although the two common models of economic impact analysis are valuable methods to estimate economic impacts after natural disasters, micro-level assessments on a small regional level are omitted from these calculations.

Specific damage assessments in the Caribbean demonstrate further ways of estimating social and economic effects of natural disasters (volcanoes, hurricanes, and earthquakes) calculated on the macro-level (Cross, 2007; Jovel, 1989). Countrywide assessments of financial damage of hurricanes David in 1979 and Hugo in 1989 were conducted by Benson, Clay, Michael, and Robertson (2001), and Benson and Clay (2004). Rasmussen (2004) evaluated the long-term macroeconomic effects of disasters from a country-level perspective highlighting the effects of fiscal balances, as well as the necessity of pre-disaster risk reduction plans essential for recovery. Aspects of recovery are visible in media portrayals of disasters suggesting damage to the entire country (Faulkner, 2001; Huang & Min, 2002), or portrayals of large financial losses by emphasizing additional negative economic impacts (Albala-Bertrand, 1993). How well the tourism sector manages to convince the public that everything has returned to normal business is seen as the short-term basis for disaster recovery (Rowe, 1996). However, natural disasters are seen often as a national problem experienced at local levels with varying impacts on the supply and demand side of the economy (Cole, 1995). This is likely to result in a weakened economic development for a region or country despite having an adaptable tourism industry (Faulkner, 2001). Although the disaster may trigger new economic growth through the stimulus of reconstruction activities, there should be adequate financing, local personnel, and materials made available (Skidmore & Toya, 2002). Hallegatte and Przyluski (2010) collected data on the global effects from the 2010 earthquake in Port-au-Prince and Hurricane Katrina 2005 to calculate direct and indirect losses, changes in housing prices, length of reconstruction phases, and the stimulus effect of a disaster that is triggered by reconstruction activities. Pelling et al. (2002) further recognized the necessity of integrating disaster vulnerability into pre-disaster development planning in assessing direct financial and indirect damages by applying a holistic accounting of macroeconomic impacts with multiple disaster case studies.

Sector-specific research and micro-level analyses of financial disaster effects are not the main focus in natural disaster assessments. Instead, much of the analysis and focus addressed the problems of landslides (DeGraff, James, & Breheny, 2010; DeGraff, Romesburg, Ahmad, & McCalpin, 2012; Maharaj, 1993). The current focus on macro-economic damage assessment in research should include examinations of individual economic sectors that were impacted by a disaster. Large differences among geographic regions and the shortage of available data at the micro-level are key reasons why existing research on natural disaster impacts seldom addresses damages and consequences at such small scales. Analyses lack a micro-level perspective for post-disaster impact analysis in these studies to provide a more in-depth critique of disaster recovery and impacts on financial aspects, particularly in the tourism industry. This implies real danger of distorting and misunderstanding disaster impacts.

Disaster impact analysis can be incorporated in various disaster assessment frameworks currently in the literature with small-scale disaster assessment in mind (Faulkner, 2001; Hystad & Keller, 2008; Lindell, Prater, Perry, & Nicholson, 2006). This should be simultaneously examined as part of sustainable disaster recovery planning, especially at the community-level given that there are significant vulnerability variations impacting potential disaster recovery (Lindell, 2011). Disaster recovery processes involve different aspects of physical, social, economic, and environmental elements for communities through pre and post-event (Smith & Wenger, 2007). Pre-event and post-event planning both are an integral part of Faulkner's (2001) tourism disaster management framework and Hystad and Keller's (2008) disaster management framework. Lindell et al. (2006) developed a disaster impact model emphasizing three emergency management interventions for the pre-disaster phase (hazard mitigation, emergency preparedness, and recovery preparedness) and a post-disaster phase (mitigation of physical and social impacts). The role of individual stakeholders is particularly emphasized by Hystad and Keller (2008). They show when stakeholders are involved in the post-disaster resolution phase then they are more likely to be better prepared for future disasters. Pre- and post-disaster recovery planning are important elements of disaster recovery theory for sustainable communities (Smith & Wenger, 2007). Often the tourism industry tends to be ill prepared for disasters regardless of having knowledge of existing vulnerabilities (Becken & Hughey, 2013). Therefore, we focus on a post-disaster impact assessment at the local-scale to support sustainable disaster recovery in order to reduce vulnerabilities for the tourism sector.

Methodology

Study area

We selected the island of Dominica as study area specific to tourism disaster recovery for three reasons. First, tourism plays an important role for island destinations resulting in economic dependency where tourist destinations are particularly susceptible to the consequences of climate change (Becken & Hay, 2007). Second, the SIDS Dominica and according tourism sector will continue to face severe challenges and consequences due to climate change since it poses the most threat for island communities (Forbes et al., 2013). Third, islands can be regarded as isolated systems, allowing measuring all economic input and output flows including the tourism industry.

Dominica is located within the Lesser Antilles in the Caribbean. Dominica is characterized by a mountainous interior covered in tropical rainforest, and a variegated coastline. Formerly reliant on the single export commodity bananas (Payne, 2008), the island is transitioning towards a promising tourism development. Dominica's Tourism Master Plan sees tourism as the main income generator of the future, drawing attention to the importance of sustainable development (Commonwealth of Dominica, 2013). Accordingly, the island is advertised as the 'nature island' boasting 'sustainable tourism' (DDA, 2013); a claim made plausible by the absence of mass tourism due to the island's geography (Timms & Conway, 2012).

Dominica's disaster vulnerability results from extensive zones of weakened rock, over steepened slopes, large rainfall amounts and occasional seismic activity that facilitate landslides (Teeuw, Rust, Solana, & Dewdney, 2009), and one of the highest concentrations of potentially active volcanoes in the world (Lindsay, Smith, Roobol, & Stasiuk, 2005). Due to its location in the tropical hurricane belt, each year hurricanes and tropical storms pass over the entire region, each one affecting different islands and to various extents. From 1872 to 2015, Dominica was hit by 22 hurricanes and 32 tropical storms (Hurricane City, 2016).

Tropical storm Erika

On 27 August 2015, tropical storm Erika passed over Dominica bringing extraordinary rainfall resulting in rapid flooding and landslides throughout the island, particularly affecting the south and southeast parts of the island (Commonwealth of Dominica, 2015). Tropical storm Erika reached peak wind speeds of 50 mph, within the wind speed range of tropical storms according to the Saffir Simpson scale (39–73 mph; National Hurricane Center, 2016). Each location of Dominica's climate stations recorded more than 200 mm of precipitation within four hours, while some areas reached a maximum of 400 mm in the same period. The rainfall peak occurred between 4 AM and 10 AM while most residents were still in their homes when the series of flashfloods and landslides began. The situation was further aggravated by surface runoff from steep interior parts of the mountainous island reaching the coastal areas within just a few hours. The island's steep water catchments faced quick and intense runoff, exacerbating the effects of the rainfall and together resulting in severe impacts for just a few areas.

Officials claimed Dominica had been set back 20 years in tourism development as a result of tropical storm Erika with nine communities declared 'special disaster areas' by Prime Minister Roosevelt Skerrit (Commonwealth of Dominica, 2015). The storm resulted in the death of twelve people, 20 injured, 22 persons were listed as missing, 574 as homeless, 713 had been evacuated, and 7229 were living in a disaster area (IFRC, 2015). Storm Erika generated total financial damages of US$ 482.84 million, corresponding to 90% of Dominica's GDP in 2015 (GFDRR, 2015). The tourism sector was the fourth highest affected sector with total financial damages of US$ 31.18 million (Table 1), resulting in severe

Table 1. Storm Erika damages and losses by sector.

Sector	Damage (US$ millions)	Loss (US$ millions)	Total (US$ millions)
Roads and bridges	239.45	48.28	287.53
Housing	44.53	9.61	54.15
Agriculture, fisheries, forestry	42.46	4.87	47.33
Tourism	19.48	11.70	31.18
Water and sanitation	17.14	2.38	19.52
Others	16.38	2.08	18.36
Air and sea ports	14.90	0.08	14.98
Industry and commerce	9.13	0.56	9.69
Total	403.28	79.56	482.84

Source: Recreated by authors using GFDRR (2015).

Table 2. Dominica change in over-night arrivals by accommodation type, May 2015 and May 2016.

Accommodation type	Over-night arrivals May 2015	Over-night arrivals May 2016	Change May 2015 vs. May 2016
Bed & Breakfast	79	32	−59.5%
Guest house	406	231	−43.1%
Hotel	1730	1057	−38.9%
Apartment/cottages	655	438	−33.1%
Dive/eco lodge	174	194	11.5%

Source: Statistical Office Dominica (2016).

damage to the national economy due to its dependency on tourism. Furthermore, indirect costs for tourism were caused by impacts to roads and bridges, further aggravating the island's tourism sector since it is highly dependent on a functioning infrastructure to access individual tourist areas.

Among the island's 95 hotels, 31 suffered direct losses, including 11 hotels that ceased operations, and two hotels were completely destroyed. The hotel losses alone add up to a financial damages of US$ 15 million for Dominica (GFDRR, 2015). As a consequence, the number of overnight arrivals decreased from 7097 in May 2015 to 5645 in May 2016, corresponding to a net loss of 20.5% (Statistical Office Dominica, 2016). Over-night visitors comparing May 2015 and May 2016, decreased for all types of accommodation within a range of 30% to 60% (Table 2).

Apart from direct effects, one example of secondary effects of the storm was the cancellation of the Creole Music Festival in October 2015, which added to the additional loss of revenue for the tourism sector. This can be illustrated by the number of over-night arrivals in the fourth quarter of the years 2013–2015 (Table 3). In 2015, the fourth quarter saw a decrease in over-night arrivals by 29.3% in comparison to 2014. For the festival month of October, there was a 46.3% decrease between the same years. Influence of the

Table 3. Dominica quarterly over-night arrivals, 2013, 2014, and 2015.

Quarterly arrivals	2015	2014	2013	Change 2015 vs. 2014
(1) Quarter	20,695	20,470	20,334	1.1%
(2) Quarter	18,211	18,614	16,622	−2.2%
(3) Quarter	20,690	21,372	20,407	−3.2%
(4) Quarter	14,878	21,055	20,914	−29.3%
-October	4611	8584	8982	−46.3%
Total	74,474	81,511	78,277	−8.6%

Source: Jacob (2016).

Creole Music Festival on tourism arrivals is also illustrated by a monthly comparison of stop-over arrivals. In October 2014, the number of overnight arrivals was 8584, exceeding the arrivals during the season in January and February (6422 and 7400 arrivals, respectively). Consequently, the drastic decrease in overnight arrivals in the fourth quarter in 2015 compared to 2014 can be attributed to the cancellation of the festival, yet it cannot be determined whether arrivals statistics have recovered or will recover in the long-term.

Furthermore, the decrease of arrivals can partly be attributed to the destruction of several accommodation facilities. The Jungle Bay Resort ('JBR') was one facility located within the community of Petite Savanne, and provided a substantial source of income for several communities in Dominica (Figure 1). This is in large part due to the accessibility of JBR from the northern parts of the island, which do not allow for employees to have long-distance commutes outside the community. As the fourth largest hotel in Dominica by room numbers, JBR displayed higher personnel expenditure than other hotels of comparable or larger size due to the focus on sustainable tourism as a luxury resort. JBR will be used as case example to illustrate the disaster impact assessment model which can be used to estimate direct and indirect economic effects of natural hazards on tourism and regional development.

Historically, the area in which JBR was located has been devoid of landslides and severe hurricane activity. Unpredictable landslides following intense rainfall caused complete destruction of the resort facilities and major flooding carrying debris cut-off the main road connecting the southwestern and southeastern parts of Dominica. To date, this break in the main road is irreparable. The communities most severely impacted (e.g. Petite Savanne, Delices, and Boetica) were unprepared for a tropical storm with such devastating destruction. Ultimately, JBR's destruction led to severe economic consequences for areas with a large share of employees or suppliers because JBR cannot be rebuilt in the same area due to its status location classified as a permanent disaster area. JBR was the only large employer in this area, and facilitated a detailed assessment of communities' job markets without being influenced by many other economic factors.

Research model and data

The tourism disaster management framework proposed by Faulkner (2001) is used as the theoretical background. The relevance and application of Faulkner's (2001) framework for this study is due to the specific focus on destinations' tourism sector based on the literature review of relevant frameworks. Faulkner (2001) emphasizes the importance of scale, specifically community scale in describing disaster stages regarding disaster response. Community scale analysis is needed because of how disaster impacts and recovery differ among tourism regions and stakeholders (Faulkner, 2001). Ritchie (2004) expands on this framework, stressing disaster recovery is a crucial aspect within the tourism industry, but there is also a need for local level approaches to test and verify models utilized in the tourism disaster management.

The framework addresses six phases of a disaster requiring different types of tourism disaster strategies:

(1) Pre-event: strategies to mitigate or prevent disaster effects.
(2) Prodromal: strategies needed when a disaster is imminent.

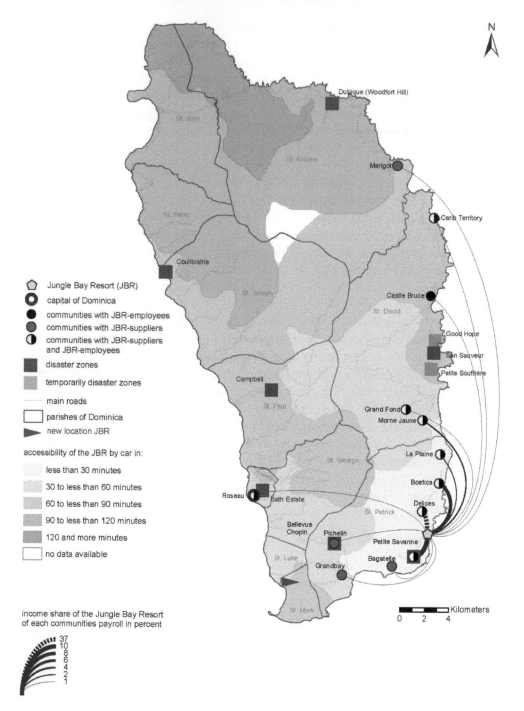

Figure 1. Research area and JBR effects on the regional labor market. Source: Designed by authors according to GFDRR (2015) and JBR (2015); database: Diva GIS.

(3) Emergency: action needed during the effect of a disaster.
(4) Intermediate: short-term measures.
(5) Long-term (recovery): continuation of phase (4).
(6) Resolution: restoring routine.

Mair, Ritchie, and Walters (2014) note that there have been few cases where Faulkner's (2001) tourism disaster framework has been tested and suggest that there is a need for detailed models at each phase of the crisis or disaster lifecycle (e.g. reduction, readiness, response, and recovery). Due to this necessity of micro-level post-disaster damage assessment in the tourism sector, we focus on the precise handling of phases (4) and (5) to facilitate a successful and time-efficient phase 6 of resolution for Faulkner's (2001) framework.

Generally, disaster impacts can be investigated focusing on aspects such as material destruction causing direct economic loss (e.g. houses, infrastructure), disruptions of social structures (e.g. family linkages) or indirect economic consequences (e.g. loss in income). This model focuses on the economic consequences of the material destruction of touristic infrastructure accompanied by a loss in income in a community. Ritchie (2004) stresses the importance of the tourism sector and disaster management needing to understand the interdependence and impacts on economies and livelihoods. This aspect is particularly relevant since this can be an important indicator of livelihoods or adequate standards of living when estimating disaster impacts due to the loss of income affecting a community (e.g. indirect impacts due to the decline in revenue in the community or migration forced by the loss of employment). However, financial compensations are generally paid based on the material destructions, which do not cover the actual economic impacts of a disaster on a community. These indirect economic consequences so far have not been investigated on a small spatial level since data on the loss of jobs in most cases is only available on a regional or national level. Therefore, such information would be crucial for the development of adequate disaster recovery strategies. Calculations of monetary damages alone, even if regionally differentiated do not provide adequate information about the economic disaster aftermath for local or national recovery programs for residents. It is necessary to measure micro-level impacts on local economic situations for disaster areas in order to allocate appropriate financial programs and implement adequate recovery programs supported by local and national governments as suggested by Pelling et al. (2002) and Skidmore and Toya (2002). Areas where large parts of the job market and income sources for residents have been terminated or reduced from disasters require more financial assistance than areas with limited economic impact relying on the reconstruction of tourism infrastructure or the tourism economy.

For these reasons we developed an economic calculation model capable of analyzing natural disaster damage in a specific localized region to obtain calculations of losses for communities managing recovery. The calculation of economic impacts in this research model is based on three spatial levels (local, regional and national) in relation to the pre-disaster touristic infrastructure incorporating data such as employment rates on country-wide and community-based levels, as well as community-by-community aggregated income. Detailed data about the precise income structures of one major tourism employer which has been destroyed during the disaster is used to calculate direct financial consequences for the corresponding dependent communities. The pre-destruction share of income of employees in specific areas provides a better understanding of the impact to the regions' job market and economic state for the residents.

Micro-level impact assessment

MLAM is calculated on the basis of data on the countrywide (Dominica), community-based and at a specific local-level (case study JBR) from the year 2015 (Figure 2) to

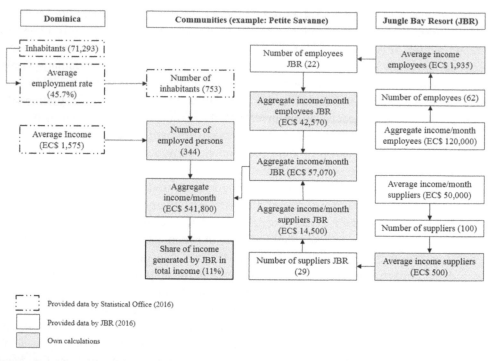

Figure 2. Micro-level analysis model (MLAM): calculation of JBR income share in total income for a sample community. Source: Formula and calculations by authors from data received by JBR (2015) and Statistical Office Dominica (2016).

estimate financial consequences for community residents formerly employed at JBR. The calculation of financial losses from JBR's destruction utilizes average numbers for Dominica's employment rate and the monthly income. Average values are used for the income of JBR employees and suppliers (e.g. room service, reservation manager, and housekeeper) and for the general level of income in Dominica, as it is not possible to determine regionally differentiated values in terms of urban versus rural areas, and job type (e.g. agricultural versus tourism sector employee), even within JBR. Petite Savanne serves as an example to illustrate the steps undertaken to calculate the share of income generated by JBR for the total income of the community. Analogously, economic impacts of employment were calculated for all communities (Table 4).

To estimate the number of employed persons in each community, Dominica's average employment rate (45.7%) was applied to the number of community residents. In 2015, 344 people were employed in Petite Savanne, 22 whom were employees of the JBR. Average incomes were calculated both countrywide and JBR-specific using average monthly wages at countrywide and local levels to determine the differences in average wage structures. The average income of EC$ 1575/month and the number of employed persons in communities were used to calculate community-wide monthly revenues of EC$ 541,800 in Petite Savanne. Dominican average monthly income (EC$ 1575/month) is below the average income for employees of JBR (EC$ 1935/month), emphasizing the significance of JBR as an employer for the region and impact on the affected communities. The average income of a JBR supplier is EC$ 500/month, resulting in aggregate incomes of JBR

Table 4. JBR community share of employees and income, 2015.

Community	Employees of JBR	Share of JBR employees	Suppliers of JBR	Share of JBR income
Delices	20	21.1%	33	37.0%
Petite Savanne	22	6.4%	29	10.5%
Boetica	4	5.9%	2	8.2%
La Plaine	6	1.2%	5	1.7%
Roseau	4	0.1%	13	0.1%
Others	6	0.1%	18	0.1%
Dominica	62	0.2%	100	0.3%

Source: Calculation by authors according to data from JBR (2015).

employees and suppliers yielding approximately EC\$ 120,000/month for JBR employees and EC\$ 50,000 for JBR suppliers. In the case of Petite Savanne, the aggregate income per month generated by JBR is EC\$ 57,070 (employees and suppliers included). This results in the share of income generated by JBR to the total income of communities (11% for Petite Savanne) (Table 4). The share of income refers to the share of income generated by suppliers and employees in the overall income of specific communities. This calculation was possible due to the availability of community-wide monthly revenues using aggregate income/month related to JBR and the aggregate income/per month for the entire community.

The resort's destruction by tropical storm Erika left 62 direct employees unemployed and immediately ceased arrangements with 100 suppliers. In 2015, 6.4% of employed people in Petite Savanne (suppliers not included) were employed by JBR. Overall income of 11% in the community was generated by employees' and suppliers' income. The highest losses of community income (employees and suppliers) are found in Delices, where 37% of income has been terminated. On a countrywide level, only 0.3% of revenue was terminated, highlighting the differences among regional losses further supporting the need to conduct analyses on a small regional scale rather than at a countrywide scale. While this share of revenue appears to be minor, the largest income shares of the JBR in community wage bills are Delices, Petite Savanne and Boetica (Figure 1). These communities were all in close proximity to JBR.

Discussion and conclusion

Financial damages of tropical storm Erika are illustrated on a countrywide level and examined as well on a smaller regional level. Commonly assessed country-level consequences of natural disasters help raise awareness of the destruction faced after a disaster. We affirm the importance of Benson et al. (2001) and Benson and Clay (2004) calling for attention to the country-level damages of disasters. Yet, we take a further step by analyzing damages at a small-scale level in order to estimate region-specific impacts as seen with of the community income loss between 2% and 37% resulting from the destruction of a significant employer. Such stark differences highlight the need for disaster funding and resources to be allocated on a community-by-community basis. The applied case study emphasizes the impacts of natural disasters are not equally shared across the island where some areas were declared disaster zones, and other areas only suffered minor damage. Therefore, a regional differentiation of financial disaster impacts is imperative for the development of effective relief measures and an adequate distribution of governmental subsidies. Calculations of impacts using MLAM can be used to identify communities with severe impacts needing immediate assistance. MLAM is a practical and innovative tool to support disaster recovery as it facilitates time-efficient assessment of direct financial disaster impacts specific to communities or local regions. The model is transferrable to other regions, can be adapted to different geographic scales (e.g. local or countrywide), and can be applied to any economic sector (e.g. manufacturing or agriculture), disaster type (e.g. earthquakes or volcanic eruptions), or any kind of industry (e.g. gastronomy or manufacturing).

Furthermore, MLAM can be replicated and adapted to pre-disaster phases, corresponding to Faulkner's (2001) phases 1 (pre-event) and 2 (prodromal) of the disaster management framework. We reaffirm the importance of the development of pre-disaster risk assessment measures for the tourism industry as outlined by Faulkner (2001), Faulkner and Vikulov (2001), Ritchie (2004) and Tsai and Chen (2011). Specifically, our micro-level assessment focuses on the facilitation of phases 4 (intermediate), 5 (long-term recovery), and 6 (resolution phase) in Faulkner's (2001) framework emphasizing the importance of post-disaster recovery. The disaster framework phases are inter-connected where post-disaster strategies are likely to impact or influence pre-disaster planning phases. Therefore, MLAM is useful in post-disaster assessment by providing valuable incentives for pre-disaster planning such as mitigation efforts, development of hypothetical disaster risk modeling, or management and distribution of financial resources for emergency services.

Despite the practical application and benefits of MLAM there are some inherent limitations of the model that must be addressed for future application or development. The model requires specific data from the labor and economic market such as income levels, employer data, supplier data, and number of employed persons to apply the model. This often presents a challenge in data acquisition due to the difficulty developing and maintaining the data where it is frequently expensive to do so for small-scale communities. In our case (JBR) we used income data calculated as averaged values on a community level as well as on an individual level. The model outcomes would have been more precise using individual income data allowing for a more in-depth small-scale impact analysis compared to other communities when assessing overall disaster losses. Additionally, having total income figures for several communities would have also provided more accurate information pertaining to loss of income. Another example would be to have the detailed supplier data for all available economic sectors (e.g. hotels, restaurants, tour operators, and shops or small business owners) in a community and not just one in our case of JBR. By capturing all representative suppliers for MLAM the result would yield a more accurate depiction of total disaster damages and losses that directly and indirectly impact the tourism sector. Lastly, MLAM presently only measures financial damages and losses at the impact of the disaster that is not necessarily a representation of real-time losses to reveal a larger time scale of losses. However, this may be resolved by performing the calculation on a regular time scale (e.g. monthly or annually).

All social impacts as well as infrastructure destruction from a disaster should be analyzed with other tools for a more comprehensive understanding and assessment of the disaster as well contributing to future research of disaster recovery. Residents of affected communities had to be relocated, and it is not known whether social structures can be transferred to the new places being created for the residents. Presently it is unclear what kind of jobs are available for former employees in the future, and how they can be reintegrated into the tourism sector, or find alternative jobs. Communities in Dominica traditionally were dependent on agriculture or partly subsistent agriculture and is now largely dependent on the tourism sector as local business owners or employees. In this new economic landscape, it is worth assessing the dangers of having a high dependency on one employer or one sector, and if tourism is the best pathway for community development. Should the tourism sector be the future pathway and leading source of GDP revenue, there needs to be more research of stakeholder's disaster recovery responses, adding on to the research of Ritchie (2004), McCool (2012), and Scolobig et al. (2014) with specific focus on the hospitality industry.

It is not enough to discuss overall economic damage alone in measuring disaster impacts. Many sectors were impacted by the disaster that incurred monetary damages. The worst sectors affected were housing, agriculture, roads, and tourism (GFDRR, 2015). Tourism is further aggravated due to the labor intensive hospitality market and dependency on other sectors such as roads and housing. Infrastructure is a critical aspect for tourism in large part due to the number of roads needed for transportation and distribution of hospitality resources. The housing sector is also important due to the availability of employee housing, but also for available accommodation units for tourists such as bed and breakfast facilities. Since the island is becoming more reliant on the tourism sector, the impact is particularly devastating for communities unable to recover after disasters. The destruction of the Jungle Bay Resort as tourism employer had far-reaching consequences for local residents, communities, and economies beyond the tourism industry. While acknowledging the work of Pelling et al. (2002), we emphasize the greater importance of assessing sectors and geographic attributes individually when natural disasters strike.

The estimation of monetary losses and localized disaster impacts by sector can serve as a foundation for focusing on long-term effects and strategies of disaster recovery for pre-disaster and post-disaster mitigation planning. A key benefit of a disaster like Tropical storm Erika is the awakening effect regarding the necessity of zoning maps and pre-disaster risk reduction plans. Awareness has been raised not only within the affected population, but also within the political and institutional systems. The phenomenon of forced adaptation should be addressed in contrast to voluntary adaptation well in advance of a disaster to assist in tourism disaster recovery to the greatest extent possible (Tervo-Kankare et al., 2016). Likewise, in the face of changing environments resulting from climate change, the focus of tourism disaster management will likely shift to use of tourism disaster management frameworks, such as Faulkner's (2001) framework utilizing different phases in the disaster process to address issues such as zoning maps, warning systems, improved communication systems and education of industry stakeholders. MLAM is a tool that we consider not confined to post-disaster assessment and recovery phases of the tourism disaster management framework. Therefore, further research should include the integration of models such as MLAM into pre-disaster planning phases, especially in the face of environmental changes such as increased hurricane frequencies or more intensive rainfall. MLAM can provide an investigative tool for pre-disaster management planning or can concentrate on other phases of Faulkner's (2001) framework to identify vulnerable communities' dependent on the tourism sector needing adequate disaster recovery planning for economic viability.

Acknowledgement

We thank Samuel Raphael and Nancy Atzenweiler from the Jungle Bay Ressort for their support and for providing data about the Jungle Bay Resort. We also thank the Editors of this Special Issue and the anonymous reviewers for their helpful comments on this article.

Disclosure statement

No potential conflict of interest was reported by the authors.

References

Albala-Bertrand, J. M. (1993). Natural disaster situations and growth: A macroeconomic model for sudden disaster impacts. *World Development, 21*, 1417–1434.

Anderson, M. B. (1995). Vulnerability to disaster and sustainable development: A general framework for assessing vulnerability. Retrieved from http://www.disaster-info.net/lideres/spanish/mexico/biblio/eng/doc6539/doc6539-a.pdf

Baade, R. A., Baumann, R., & Matheson, V. (2007). Estimating the economic impact of natural and social disasters, with an application to hurricane Katrina. *Urban Studies, 44*, 2061–2076.

Becken, S., & Hay, J. E. (2007). *Tourism and climate change. Risks and opportunities.* Clevedon; Buffalo, NY; Toronto: Channel View Publications.

Becken, S., & Hughey, K. F. D.. (2013). Linking tourism into emergency management structures to enhance disaster risk reduction. *Tourism Management, 36*, 77–85.

Becken, S., Mahon, R., Rennie, H. G., & Shakeela, A. (2014). The tourism disaster vulnerability framework: An application to tourism in small island destinations. *Natural Hazards, 71*, 955–972.

Benson, C., Clay, E., Michael, F. V., & Robertson, A. W. (2001). Dominica: Natural disasters and economic development in a small island state. Disaster Risk Management Working Paper Series No. 2. Washington, DC. Retrieved from https://www.odi.org/sites/odi.org.uk/files/odi-assets/publications-opinion-files/4792.pdf

Benson, C., & Clay, E. (2004). Understanding the economic and financial impacts of natural disaster. Disaster Risk Management Series No. 4. Washington, DC. Retrieved from ipcc-wg2.gov/njlite_download2.php?id = 9000

Briguglio, L. (2003). The vulnerability index and small island developing states. A review of conceptual and methodological issues. Retrieved from http://www.um.edu.mt/__data/assets/pdf_file/0019/44137/vulnerability_paper_sep03.pdf

Cavallo, E., Powell, A., & Becerra, O. (2010). Estimating the direct economic damages of the earthquake in Haiti. *The Economic Journal, 120*, F298–F312.

Coffman, M., & Noy, I. (2011). Hurricane Iniki: Measuring the long-term economic impact of a natural disaster using synthetic control. *Environment and Development Economics, 17*, 187–205.

Cole, S. (1995). Lifelines and livelihood: A social accounting matrix approach to calamity preparedness. *Journal of Contingencies and Crisis Management, 3*, 228–246.

Collymore, J. (2011). Disaster management in the Caribbean: Perspectives on institutional capacity reform and development. *Environmental Hazards, 10*, 6–22.

Commonwealth of Dominica. (2013). Tourism master plan 2012–2022. Final Report. Retrieved from http://tourism.gov.dm/images/documents/tourism_master_plan/tourism_master_plan_june2013.pdf

Commonwealth of Dominica. (2015). Rapid damage and impact assessment. *Tropical Storm Erika*. Retrieved from http://www.dominica.gov.dm/images/documents/rapid_damage_impact_assessment_dominica.pdf.

Cross, J. A. (2007). Natural hazards within the West Indies. *Tourism Geographies, 91*, 190–199.

DeGraff, J. V., James, A., & Breheny, P. (2010). The formation and persistence of the Matthieu Landslide-Dam Lake, Dominica, West Indies. *Environmental & Engineering Geoscience, 16*, 73–89.

DeGraff, J. V., Romesburg, H. C., Ahmad, R., & McCalpin, J. P. (2012). Producing landslide-susceptibility maps for regional planning in data-scarce regions. *Natural Hazards, 64*, 73–89.

Discover Dominica Authority (DDA). (2013). Welcome to Dominica. Retrieved from http://www.dominica.dm/index.php

Durocher, J. (1994). Recovery marketing: What to do after a natural disaster. *The Cornell Hotel and Restaurant Administration Quarterly, 35*, 66–71.

Dwyer, L., Forsyth, P., & Spurr, R. (2004). Evaluating tourism's economic effects: New and old approaches. *Tourism Management, 25*(3), 307–317.

ECLAC (Economic Commission for Latin America and the Caribbean). (2003). Handbook for estimating the socio-economic and environmental effects of disasters. Retrieved from http://eprints.mdx.ac.uk/3979/1/1099_eclachandbook.pdf

Ellson, R. W., Milliman, J. W., & Roberts, R. B. (1984). Measuring the regional economic effects of earthquakes and earthquake predictions. *Journal of Regional Science, 24*, 559–579.

Ewing, B. T., & Kruse, J. B. (2005). Hurricanes and unemployment. Retrieved from http://citeseerx.ist.psu.edu/viewdoc/download?doi = 10.1.1.694.315&rep = rep1&type = pdf

Faulkner, B. (2001). Towards a framework for tourism disaster management. *Tourism Management, 22*, 135–147.

Faulkner, B., & Vikulov, S. (2001). Katherine, washed out one day, back on track the next: A post-mortem of a tourism disaster. *Tourism Management, 22*, 331–344.

Ferdinand, I. M., Haynes, T., & Richards, M. (2014). Assessing the vulnerability and adaptive capacity of communities to hazards and climate change in SIDS. *Proceedings of the International Concerence "AdaptToClimate", Nicosia, Cyprus*. Nicosia, Cyprus: Cypadapt.

Forbes, D. L., James, T. S., Sutherland, M., & Nichols, S. E. (2013). Physical basis of coastal adaptation on tropical small islands. *Sustainability Science, 8*, 327–344.

Forster, J., Schuhmann, P. W., Lake, I. R., Watkinson, A. R., & Gill, J. A. (2012). The influence of hurricane risk on tourism destination choice in the Caribbean. *Climatic Change, 114*, 745–768.

GFDRR (Global Facility for Disaster Reduction and Recovery). (2015). *Rapid damage and impact assessment tropical storm Erika – August 27, 2015*. Roseau. Retrieved from https://www.gfdrr.org/sites/default/files/publication/Commonwealth of Dominica - Rapid Damage and Needs Assessment Final Report.pdf

Hallegatte, S., & Ghil, M. (2008). Natural disasters impacting a macroeconomic model with endogenous dynamics. *Ecological Economics, 68*, 582–592.

Hallegatte, S., & Przylusky, V. (2010). The economics of natural disasters. Concepts and methods. Policy Research Working Paper 5507. Retrieved from https://www.researchgate.net/profile/Valentin_Przyluski/publication/227345315_The_Economics_of_Natural_Disasters/links/54574e210cf2cf5164807c91.pdf

Heger, M., Julca, A., & Paddison, O. (2008). Analysing the impact of natural hazards in small economies. The Caribbean case. UNU-WIDER Research Paper No. 2008/25. Retrieved from http://cite seerx.ist.psu.edu/viewdoc/download?doi = 10.1.1.431.7701&rep = rep1&type = pdf

Horwich, G. (2000). Economic lessons of the Kobe earthquake. *Economic Development and Cultural Change, 48,* 521–542.

Huang, J.-H., & Min, J. C. H. (2002). Earthquake devastation and recovery in tourism: The Taiwan case. *Tourism Management, 23,* 145–154.

Hurricane City. (2016). Dominica's history with tropical systems. Retrieved from http://www.hurrica necity.com/city/dominica.htm

Hystad, P. W., & Keller, P. C. (2008). Towards a destination tourism disaster management framework: Long-term lessons from a forest-fire disaster. *Tourism Management, 29*(1), 151–162.

Ibarrarán, M. E., Ruth, M., Ahmad, S., & London, M. (2009). Climate change and natural disasters: Macroeconomic performance and distributional impacts. *Environment, Development and Sustainability, 11,* 549–569.

International Federation of Red Cross (IFRC) (2015). Emergency plan of action. Dominica: Tropical storm Erika. Retrieved from http://reliefweb.int/sites/reliefweb.int/files/resources/MDRDM002E PoA.pdf

Intergovernmental Panel on Climate Change (IPCC). (2014). Climate Change 2014. Impacts, Adaptation and Vulnerability. Retrieved from https://www.ipcc.ch/pdf/assessment-report/ar5/wg2/ WGIIAR5-IntegrationBrochure_FINAL.pdf

Intergovernmental Panel on Climate Change (IPCC) (2007). Climate change 2007. Impacts, adaptation and vulnerability. Retrieved from https://www.ipcc.ch/pdf/assessment-report/ar4/wg2/ ar4_wg2_full_report.pdf

Jacob, M. (2016). 2015 Visitor statistics report. Retrieved from http://tourism.gov.dm/images/docu ments/dominica_2015_visitor_report.pdf

Jovel, J. R. (1989). Natural disasters and their economic and social impact. *Cepal Review, 38,* 133–145.

Jungle Bay Resort. (2015). *Hospitality and general data.* Petite Savanne: Jungle Bay Report.

Kahn, M. E. (2005). The death toll from natural disasters: The role of income, geography, and institutions. *The Review of Economics and Statistics, 87,* 271–284.

Kelman, I., & West, J. J. (2009). Climate change and small island developing states: A critical review. *Ecological and Environmental Anthropology, 5,* 1–16.

Kim, H., & Marcouiller, D. W. (2015). Considering disaster vulnerability and resiliency: The case of hurricane effects on tourism-based economies. *Annals of Regional Science, 54,* 945–971.

Knutson, T. R., McBride, J. L., Chan, J., Emanuel, K., Holland, G., Landsea, C., ... Sugi, M. (2010). Tropical cyclones and climate change. *Nature Geoscience, 3,* 157–163.

Koks, E. E., Carrera, L., Jonkeren, O., Aerts, J. C. J. H., Husby, T. G., Thissen, M., ... Mysiak, J. (2016). Regional disaster impact analysis: Comparing input-output and computable general equilibrium models. *Natural Hazards and Earth Systems Sciences, 16*(8), 1911–1924.

Kumar, J., & Hussain, K. (2014). Evaluating tourism's economic effects: Comparison of different approaches. *Procedia, 144,* 360–365.

Lee, Y.-F., & Chi, Y.-Y. (2013). Natural disaster impact on annual visitors of recreation area: The Taiwan case. *International Journal of Social, Behavioral, Educational, Economic, Business and Industrial Engineering, 7,* 2715–2719.

Lindell, M. K. (2011). Disaster studies. *Current Sociology, 61*(5–6), 797–825.

Lindell, M. K., Prater, C. S., Perry, R. W., & Nicholson, W. C. (2006). *Fundamentals of emergency management.* Washington, DC: FEMA.

Lindsay, J. M., Smith, A. L., Roobol, M. J., & Stasiuk, M. V. (2005). Dominica. In M. Lindsay, R. E. A. Robertson, J. B. Sheperd, & S. Ali (Eds.), *Volcanic hazard atlas of the lesser Antilles* (pp. 2–48). St. Augustine: Seismic Research Centre. Retrieved from http://uwiseismic.com/downloads/dominica_vha. pdf

Mair, J., Ritchie, B. W., & Walters, G. (2014). Towards a research agenda for post-disaster and post-crisis recovery strategies for tourist destinations: A narrative review. *Current Issues in Tourism, 19,* 1–26.

Maharaj, R. J. (1993). Landslide processes and landslide susceptibility analysis from an upland water-shed: A case study from St. Andrew, Jamaica, West Indies. *Engineering Geology, 34*, 53–79.

McCool, B. N. (2012). The need to be prepared: Disaster management in the hospitality industry. *Journal of Business & Hotel Management, 1*, 1–5.

Méheux, K., Dominey-Howes, D., & Lloyd, K. (2007). Natural hazard impacts in small island development states. *Natural Hazards, 40*, 429–446.

Méheux, K., & Parker, E. (2006). Tourist sector perceptions of natural hazards in Vanuatu and the implications for a small island developing state. *Tourism Management, 27*, 69–85.

Merz, B., Kreibich, H., Thieken, A., & Schmidtke, R. (2004). Estimation uncertainty of direct monetary flood damage to buildings. *Natural Hazards and Earth System Sciences, 4*, 153–163.

Mimura, N., & Nurse, J. (2007). Small islands. Retrieved from https://www.ipcc.ch/pdf/assessment-report/ar4/wg2/ar4-wg2-chapter16.pdf

National Hurricane Center. (2016). Saffir-Simpson Hurricane wind scale. Retrieved from http://www.nhc.noaa.gov/aboutsshws.php.

Neumayer, E., Plümper, T., & Barthel, F. (2014). The political economy of natural disaster damage. *Global Environmental Change, 24*, 8–19.

Noy, I. (2009). The macroeconomic consequences of disasters. *Journal of Development Economics, 88*, 221–231.

Payne, A. (2008). After bananas: The IMF and the politics of stabilization and diversification in Dominica. *Bulletin of Latin American Research, 27*, 317–332.

Pelling, M., Özerdem, A., & Barakat, S. (2002). The macro-economic impact of disasters. *Progress in Development Studies, 2*, 283–305.

Pelling, M., & Uitto, J. I. (2001). Small island developing states: Natural disaster vulnerability and global change. *Environmental Hazards, 3*, 49–62.

Rasmussen, T. N. (2004). *Macroeconomic implications of natural disasters in the Caribbean* (Working Paper 224). Retrieved from https://www.imf.org/external/pubs/ft/wp/2004/wp04224.pdf.

Ritchie, B. W. (2004). Chaos, crises and disasters: A strategic approach to crisis management in the tourism industry. *Tourism Management, 25*, 669–683.

Rowe, M. (1996). Caribbean comeback. *Lodging Hospitality, 5*(1), 28–31.

Sadowski, N. C., & Sutter, D. (2005). Are safer hurricanes more damaging? *Southern Economic Journal, 72*, 422–432.

Scolobig, A., Komendantova, N., Patt, A., Vinchon, C., Monfort-Climent, D., Begoubou-Valerius, M., ... Di Ruocco, A. (2014). Multi-risk governance for natural hazards in Naples and Guadeloupe. *Natural Hazards, 73*, 1523–1545.

Selcuk, F., & Yeldan, E. (2001). On the macroeconomic impact of the August 1999 earthquake in Turkey: A first assessment. *Applied Economics Letters, 8*, 483–488.

Skidmore, M., & Toya, H. (2002). Do natural disasters promote long-run growth? *Economic Inquiries, 40*, 664–687.

Smith, G. P., & Wenger, D. (2007). Sustainable disaster recovery: Operationalizing an existing agenda. In H. Rodgriguez, E. Quarantelli, & R. Dynes (Eds.), *Handbook of Disaster Research* (pp. 234–257). New York, NY: Springer.

Statistical Office Dominica. (2016). Written announcement of statistical data 2015. Dominica.

Strobl, E. (2012). The economic growth impact of natural disasters in developing countries: Evidence from hurricane strikes in the Central American and Caribbean regions. *Journal of Development Economics, 97*, 130–141.

Teeuw, R., Rust, D., Solana, C., & Dewdney, C. (2009). Large coastal landslides and tsunami hazard in the Caribbean. EOS, transactions. *American Geophysical Union, 90*, 81–82.

Tervo-Kankare, K., Kajan, E., & Saarinen, J. (2016, October 3). Costs and benefits of environmental change in arctic tourism: The industry's perceptions and responses to changes in northern Finland. Address at the meeting of Tourism Naturally Conference, Alghero, Italy.

Timms, B. F., & Conway, D. (2012). Slow tourism at the Caribbean's geographical margins. *Tourism Geographies, 14*, 396–418.

Toya, H., & Skidmore, M. (2005). *Economic development and the impacts of natural disasters* (Working Paper 05-04). Retrieved from http://citeseerx.ist.psu.edu/viewdoc/download?doi=10.1.1.660.1931&rep = rep1&type = pdf.

Trenberth, K. E. (2011). Changes in precipitation with climate change. *Climate Research, 47*, 123–138.

Tsai, C.-H., & Chen, C.-W. (2011). The establishment of a rapid natural disaster risk assessment model for the tourism industry. *Tourism Management, 32*, 158–171.

Tsai, C.-H., Wu, T.-C., Wall, G., & Linliu, S.-C. (2016). Perceptions of tourism impacts and community resilience to natural disasters. *Tourism Geographies, 18*, 152–173.

Tsao, C.-Y., & Ni, C.-C. (2016). Vulnerability, resilience, and the adaptive cycle in a crisis-prone tourism community. *Tourism Geographies, 18*, 80–105.

Vigdor, J. (2008). The economic aftermath of hurricane Katrina. *Journal of Economic Perspectives, 22*, 135–154.

West, C. T., & Lenze, D. G. (1994). Modeling the regional impact of natural disaster and recovery: A general framework and an application to hurricane Andrew. *International Regional Science Review, 17*, 121–150.

Exploring stakeholder groups through a testimony analysis on the Hawaiian aquarium trade

Brooke A. Porter ⓘ

ABSTRACT

A remote archipelago, Hawai'i, offers a plethora of sought after coastal and marine tourism experiences. The same unique marine fauna that draws tourists also makes Hawai'i a major player in the international ornamental aquarium trade. For many residents of Hawai'i, the marine realm is part of their island home and interactions with tourists and tourism activities are a part of everyday life. For many residents, the ocean is an important resource and for some a staple source of livelihood, be it through tourism, fisheries, or the aquarium trade. This variance between extractive and non-extractive marine resource use creates conflicts between stakeholder groups in Hawai'i. This study thematically analyzes public testimony records, which included 1652 individual testimonies, from proposed legislation aimed to establish 'an aquatic life conservation program in the Division of Aquatic Resources to implement conservation measures, including limited entry areas and certification requirements, to regulate the collection of fish and other aquatic life for aquarium purposes. Public opinions evident in individual testimonies are representative of the disagreement in the literature regarding the stability and health of reef fishes populations in Hawai'i and broader resource-user conflicts. This study aims to better describe the user conflict between stakeholder groups in the marine resources of Hawai'i by exposing themes concerning change in natural environments.

摘要

夏威夷这个边远群岛不仅能够提供各种寻求海岸与海洋旅游体验的机会, 同样吸引旅游者的还有其海洋生物群落, 从而使其成为国际装饰用水生物贸易的重要一员。对很多居民来说, 海洋领域是他们岛屿家园的一部分, 与旅游者交往, 参与各种旅游活动也是其日常生活的一部分。对很多居民来说, 海洋是一种重要的资源, 对另一些人来说, 不管从事旅游、渔业还是水生物贸易的居民, 海洋还是不可或缺的生计来源。这种消耗性与非消耗性海洋资源使用的差异造成夏威夷利益群体之间的冲突。本研究从理论上分析了公众的证词记录, 从水资源部门为了保护水生物项目的建议立法(包括限制进入区域、认证许可) 到对以水牛物贸易日的采集鱼类及其它水牛物资源进行管冶(夏威夷2015年 873号法案)。在公民个别证词里面表现出的看法反映了文献中对保护夏威夷礁石鱼类种群稳定与健康以及广泛意义上的资源使用者冲突。本文通过曝光自然资源环境变化议题旨在较好地描绘夏威夷海洋资源利益群体的资源使用冲突。

Introduction

Hawai'i is known by tourists as the 'Paradise of the Pacific' (see Mak, 2015). Yet within the visitor and marine tourism industries in Hawai'i, comments and concerns regarding a reduction in reef fish are commonplace (personal observation, 2001–2010). Previous research has found that tourist perceptions of degraded coastal resources are often comparable to actual ecological impacts documented in the literature (Priskin, 2003). It is evident that aquarium collectors and many marine tour operators (including divers, snorkelers, glass-bottom boats, and submarines) are in competition for the same resources – vibrant, colorful, and even rare or endemic marine organisms. In Hawai'i, there is disagreement in the literature on the status and health of many fish populations. While some studies have shown some fish populations to be increasing in fish replenishment areas (FRAs) (e.g. Tissot, Walsh, & Hixon, 2009; Walsh, 2015; Walsh, Cotton, Dierking, & Williams, 2003), a recent NOAA report acknowledges 'the relatively data-poor status of most of the coral-reef fish stocks in Hawai'i' (Nadon, 2017, p. 3). Despite this direct conflict of interests, there is a paucity in the literature on the resource-user conflict between the aquarium trade and the tourism industry, as well as on coastal marine resource conflict in general (Hannak, Kompatscher, Stachowitsch, & Herler, 2011; Voyer, Barclay, McIlgorm, & Mazur, 2017).

The interconnectedness of marine resource use is obvious, yet, struggle for control of resources creates conflicts of interest among marine resource stakeholders and users (Christie, 2004; Majanen, 2007). For some, marine and coastal resources provide a sought after recreational escape. For others, marine aquaria are sought out for entertainment or enjoyment purposes (Packer & Ballantyne, 2002). The user conflict associated with aquarium collection and other non-extractive uses such as tourism and recreation is based on the circumstance that one user benefits from the permanent removal of the marine resource, while the other user benefits from the continuous viewing of the marine resource in its natural habitat and environment (e.g. Tissot, 2005; Tissot et al., 2009). Many forms of marine tourism rely upon the viewing of the marine resources (e.g. SCUBA diving, snorkeling); however, there are extractive uses of the marine resource associated with marine and coastal tourism (e.g. sport fishing, seafood consumption, shell curio) (Voyer et al., 2017). Marine tourism, like any other industry, carries its own environmental risks, including, but not limited to reef trampling, pollution, and effluent (e.g. Ong, Storey, & Minnery, 2011). Additionally, coastal and marine tourism can increase the demand for seafood (Voyer et al., 2017). Intersectoral conflict between tourism and fisheries has been documented in the food fisheries literature (e.g. Fabinyi, 2008; Ong et al., 2011). In places like Hawai'i where tourism, and more specifically marine tourism, is an important economic sector, resource-use conflict and stakeholder perceptions must be carefully considered.

The aquarium industry has been criticized for its potential environmental risks to marine ecosystems. Invasive species and the introduction of disease have resulted from the aquarium industry (Rhyne et al., 2012), with the case of the lionfish (*Pterois volitans*) in the Atlantic as a prominent example (Albins & Hixon, 2013). Anecdotal evidence from tourists and residents suggests an associated decline in reef populations; however, the direct environmental impacts of the aquarium trade to coral reefs ecosystems are not well known (see also Nadon, 2017). Population assessments do not exist for at least 64% of the globally traded species (Rhyne et al., 2012).

Hawai'i is a major supplier for the aquarium trade (Tissot & Hallacher, 2003; Walsh et al., 2003). This is of potential concern as Hawai'i's collection restrictions are minimal, and maximum sustainable yields for Hawaiian ornamentals are lacking (Ogawa & Brown, 2001). Further, the accuracy of ornamental catch reporting in Hawai'i is questionable (Rhyne et al., 2012; Tissot & Hallacher, 2003; Walsh, 2000) and can sometimes go undocumented as a result of legal loopholes regarding dealer reporting (Walsh, 2015). The Department of Land and Natural Resources–Division of Aquatic Resources (DLNR-DAR) is the state agency in charge of aquarium collection permitting. There are two types of permits issued by DLNR-DAR: commercial and non-commercial. Persons holding a commercial collector permit are required to pay an annual fee (USD $50) and report catches; there are no fees or reporting requirements for non-commercial permit holders. The reporting regulations on aquarium dealers is negligible (Walsh, 2015) complicating the traceability of catches and creating a potential for undocumented sales.

There is some evidence indicating the aquarium trade has impacted Hawai'i fish populations. For example, the Achilles tang (*Acanthurus achilles*), listed as an ecologically unsustainable species by the Sustainable Aquarium Industry Association (SAIA), is the third most collected species in Hawai'i for aquaria (Walsh, 2015); larger specimens are also taken in Hawai'i for sustenance purposes. In addition to aquarium collection, there are a number of other issues impacting the marine environments of Hawai'i (e.g. injection wells, non-point source pollution, climate change, reef trampling), some of which are associated with tourism. While these issues are also important, the pronounced user conflict between recreation/tourism and aquarium collectors is the subject of this analysis.

Rationale

Tourism is an important economic sector for Hawai'i and among the major state sectors, it supports the largest number of jobs statewide (HTA, 2016). With Hawai'i acting a major source of marine organisms for the aquarium trade, specifically in the USA (Hawaii State Legislature, 2015; Tissot & Hallacher, 2003; Walsh et al., 2003), public concern for the impact of the aquarium trade on the state's reef ecosystem began as early as 1973 (Walsh, 2000). A 2012 Hawai'i survey of residents showed that two-thirds of the sample were in favor of a full ban on aquarium fish collection (Muller, 2012). A 2017 poll showed that the numbers in favor are now as high as 90% (Big Island Now.Com, 2017). Given the economic importance of the tourism industry to Hawai'i and the significant role the state's resource plays in the aquarium trade, understanding intersector challenges resulting from direct conflict of resource use is critical for the state. With the exception of Tissot's (2005) Integral Ecology (IE) analysis of the West Hawai'i Fisheries Council, there has been a lack of qualitative research efforts to understand stakeholder perceptions surrounding the ornamental aquarium trade. Therefore, this study aims to build upon the previous research by identifying emergent themes in public testimony records between marine user groups in support of and in opposition to the ornamental aquarium trade in Hawai'i using an exploratory thematic testimony analysis from proposed Hawaii House Bill 873 (hereon referred to as HB 873) (see Hawaii State Legislature, 2015). HB 873 (Hawaii State Legislature, 2015), available in full online, seeks to establish an aquatic life conservation program in the Division of

Aquatic Resources to implement conservation measures, including limited entry areas and certification requirements, to regulate the collection of fish and other aquatic life for aquarium purposes (Hawaii State Legislature, 2015). The bill was founded in the belief that collection of marine organisms for the aquarium trade poses a significant environmental and socioeconomic threat to the state, that the trade has operated for over 50 years without take or permit limits, and that the industry targets herbivorous species which are considered critical to reef health. Specific goals of HB 873 are to ensure sustainability of the state's fisheries, protect animal welfare, increase regulations (e.g. bag limits) for the aquarium trade, respect cultural rights, avoid interference with subsistence fishing, and reduce fatality rates to 1% or less for those organisms targeted by the marine aquarium trade. If passed and implemented, HB 873 has the potential to impact multiple user groups (e.g. aquarium collectors, aquarium hobbyists, tour operators, tourists, recreationists). Given the complexities of researching multiple user groups, public hearing sometimes is 'the only form of interaction between the agency and the affected public' (Fiorino, 1990, p. 230). Stepanova (2015) notes that public forums associated with the introduction of new information (e.g. a bill) often reveal areas of competition and conflict surrounding the issue. Further, public testimony introduces important lay judgments and observations (see Fiorinio, 1990). Similar to Froehlich, Gentry, Rust, Grimm, and Halpern (2017), this research assumes that the direct and persuasive language characteristic to public comments is 'ideal for analyzing sentiment polarity' (Government public comments, para. 1). Thus, the exploration of public testimony on HB 873 provides an important opportunity to explore opinions and potential areas of conflict between resource-user groups.

Methods

Public testimonies submitted in response to HB 873 were used as qualitative data. Testimonies were submitted in person, online via standard electronic form, or through email. The electronic form provided individuals with the option to oppose, support, or offer only comments on HB 873. There were two forms associated with electronic submission. One form required individuals submitting testimony via the online portal were required to enter a name, email, and asked to enter island or city, state, and message to the committee, while the second form required name, organization (if any), testifier position (support, oppose, comments only), and presence or absence at hearing. In some cases, individuals provided additional information including, but not limited to occupation. Testimonies are available as public record and can be found online as downloadable PDF files (see Hawaii State Legislature, 2015). Two separate hearings were held for HB 873, the first on 11 February 2015, and the second hearing on 27 February 2015. All 1652 individual records from both hearings submitted for HB 873 were used as the data-set for this study. Demographics (e.g. name, place of residence) and testimonies were entered by the author into an excel database.

Records from individuals who gave testimony on both dates of the HB 873 hearings were entered as separate testimonies. It is important to note that the majority of individuals held the same position at both hearings; however, there were two cases in which individuals changed position (both changing from opposition to support of the bill). There were multiple cases of individuals providing duplicate, triplicate, quadruplicate, or even quintuplicate testimonies for a single hearing date. Names were used to identify these cases, and duplicate testimonies were deleted so that only one testimony per person per date was allowed. In cases in which individuals submitted non-duplicate multiple testimonies, the entries were combined to form a single representative entry per person, per date. In total, 197 records were deleted from the data-set leaving a total of 1455 records as viable data for thematic analysis.

Latent thematic analysis

Analyses of testimonies are often associated with the Not in My BackYard (NIMBY) construct. However, in this case, both those in support of the bill and those opposed to the bill are in competition for the same resource – the ocean, thus, limiting the applicability of a NIMBY analysis. Wolsink (2006) argued that a NIMBY analysis can impede understanding of conflict. Further, NIMBY was not entirely applicable to the case of the aquarium trade as many opinions addressed the issue beyond a local level. Nonetheless, common NIMBY themes including organizational distrust, access to information, risk aversion, localization of the issues, and emotional attachments were considered relevant to the data and were considered in the analysis and discussion. To ensure public opinions and perceptions were adequately revealed in the data, a latent thematic analysis was used for this study. Subsections of the data including spatial demographics and general position on HB 873 were quantified.

Latent thematic analysis requires that the data-sets be organized through a coding system which divides, subdivides, and categorizes the data (Bell, 2005; Braun & Clarke, 2006; Fereday & Muir-Cochrane, 2006; Miles & Huberman, 1994). At the latent level, a thematic analysis places value on the collective meaning of the words, in this case public opinion, rather than placing meaning on a single word and in doing so, begins to identify 'underlying ideas, assumptions, and conceptualizations - and ideologies - that are theorized as shaping or informing the semantic content of the data' (Braun & Clarke, 2006, p. 13). When set in a constructivist paradigm, as was the case in this research, a latent thematic analysis overlaps with a thematic discourse analysis (Braun & Clarke, 2006). Data were initially coded electronically using TAMS Analyzer, an open source coding software; however, the individual testimonies were in multiple document formats including Microsoft Word Docs, electronic submission forms, and scans of handwritten testimonies. The various formats created issues with importing the data-sets. In addition, issues of data retrieval were becoming noticeable with the TAMS interface similar to those described by Welsch (2002). Nearly one-fifth of the data were coded electronically before shifting to manual coding using Excel datasheets. Although, there is a rigor associated with electronic coding and a potential reduction in human error (Welsch, 2002), the segmenting, coding, and collation of the data are still left up to the user (Basit, 2010). Using an inductive or constructivist approach set in a realist paradigm, the analysis followed the six phases of thematic analysis as outlined by Braun and Clarke (2006) (see Table 1).

Ethics

The testimonies used for this study is part of public record and are openly accessible as data (see Spicker, 2007). However, the participants, in this case those submitting testimony, did not consent to the research process. Therefore, participant anonymity was prioritized and names were withheld.

Table 1. Phases of thematic analysis.

Phase	Description of the process
1. Familiarizing yourself with your data:	Transcribing data (if necessary), reading and re-reading the data, noting down initial ideas.
2. Generating initial codes:	Coding interesting features of the data in a systematic fashion across the entire data-set, collating data relevant to each code.
3. Searching for themes:	Collating codes into potential themes, gathering all data relevant to each potential theme.
4. Reviewing themes:	Checking if the themes work in relation to the coded extracts (Level 1) and the entire data-set (Level 2), generating a thematic 'map' of the analysis.
5. Defining and naming themes:	Ongoing analysis to refine the specifics of each theme, and the overall story the analysis tells, generating clear definitions and names for each theme.
6. Producing the report	The final opportunity for analysis. Selection of vivid, compelling extract examples, final analysis of selected extracts, relating back of the analysis to the research question and literature, producing a scholarly report of the analysis.

Taken from Braun and Clarke (2006, p. 87).

Results

Data from 1455 testimonies were collated under two positions relating to HB 873 'opposition' or 'support.' Those in opposition to HB 873 did not want to see the aquarium trade further regulated. Those in support, favored increase regulations or bans on the aquarium trade. Thirteen individuals submitted only comments on HB 873; however, in each case of comments, it was possible to code testimonies under the two overarching positions of opposition or support. For example, it was typical for an individual submitting only comments to literally state support or opposition in their comments. In total a slight majority of individuals held opposition to HB 873 ($n = 769\%$ or 53%), with 686 (47%) in support. There were significant differences in the spatial demographics between the two groups (see Table 2).

Regions as defined by the United States Census Bureau were used to classify spatial data (see www.census.gov). Testimonies from residents of Hawai'i, normally part of the USA West region, were separated to identify local participation in the issue. For both the opposition and supporters, the majority of each group's testimonies came from Hawai'i (62% and 52%, respectively). While only 4% of the opposition's testimonies came from the mainland USA, 30% of testimonies from supporters originated from mainland USA. Likewise, there was a significant difference in the numbers of global testimonies, with the supporters having a notably higher percentage of global participation (refer to Table 2).

Table 2. Spatial demographics of individuals providing public testimony.

Region	Opposition	Supporters
USA Hawai'i	478 (62%)	358 (52%)
USA West (excluding Hawai'i)	12	118
USA Midwest	5	30
USA South	16	36
USA Northeast	1	18
USA (excluding Hawai'i) TOTAL	34 (4%)	202 (30%)
North America (excluding USA)	0	13
Europe	7	36
Africa	0	2
Asia	0	1
Oceania	0	12
GLOBAL TOTAL	7 (1%)	64 (9%)
Not specified	250 (33%)	62 (9%)

Opposition testimony came from three countries, while support testimony came from 15 different countries.

Emergent themes

The following 10 themes emerged from the data: sustainability, regulations, legislation (bill quality), economy, user conflict, social benefit, environment, animal welfare, extractive resource use, and culture (see Figure 1).

Various organizations and groups distributed prefabricated testimonies via listserv, newsletter, websites, or other outlets. Many individuals submitting testimonies used these prefabricated 'templates' as testimonies, while some combined template testimonies with additional individual comments. The use of template testimonies had significant influence on the strength of multiple themes as seen in Figure 1. Template testimonies were kept as part of the data due in part to the fact that many individuals utilizing template testimony added additional comments, it was thought that the templates chosen were representative and in support of the individuals' beliefs toward HB 873.

Theme overlap between the supporters and opposition was evident in 9 out of 10 themes; however, within each emergent theme, the subthemes of the opposition and supporters diverged. Themes and subthemes are presented in Table 3 alongside overall theme rank.

Table 3 presents a ranking of themes. For example, sustainability was the most mentioned theme by the opposition and the third most mentioned theme by the supporters. Ranks in common have been highlighted in grey. The numbers presented in parentheses

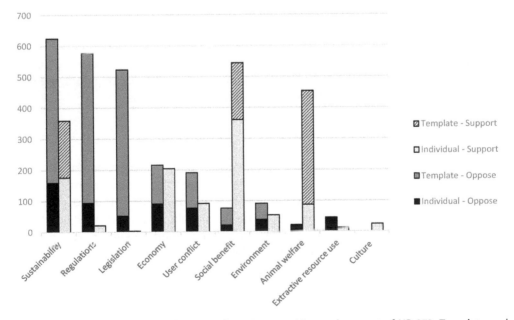

Figure 1. Comparison of themes between those in opposition and support of HB 873. Template and individual testimonies are presented as stacked columns to compare total opposition and support for the bill. Template testimonies accounted for the majority of the oppositions statements.

Table 3. Subthemes from public testimony on HB 873 (HI Rev Stat § 711-1109, 2015).

Rank	Opposition subthemes	Emergent theme	Supporters subthemes	Rank
1	Sustainable (447)	*Sustainability*	Noticeable decline in fish populations (278)	3
	Scientific support (277)		Unsustainable (61)	
			Become a model state (7)	
			Reformed collector (6)	
			Require EIS (4)	
			More MPAs needed (7)	
			Needs more research (3)	
2	Sufficient management (527)	*Regulations*	Ineffective management (21)	7
	Leave it to the experts (49)			
3	Poorly worded bill (382)	*Legislation*	Confrontational language (3)	10
	Confrontational language (77)			
	Coral not harvested (64)			
4	Negative impact on livelihoods & business (196)	*Economy*	Value of tourism (115)	4
	Supports tourism (20)		Diver/visitor commentary (89)	
5	Choosing an industry (90)	*User conflict*	Greed (61)	5
	Emotional response (70)		Fish as ornaments (28)	
	Personal enjoyment from hobby (28)		Violence (3)	
	Collection outside tourists' view (3)			
6	Concern for reef health (69)	*Social benefit*	Collective resources (268)	1
	Education/last hope (6)		Future generations (220)	
	"Collective resources" (2)		Unethical (54)	
			Public aquaria only (2)	
7	Other environmental issues more pressing (91)	*Environment*	Complex ecosystem (32)	6
			Other environmental issues (21)	
8	Best practices (21)	*Animal welfare*	High mortality (233)	2
			Inhumane (220)	
9	Don't ban fishing (32)	*Extractive resource use*	Prioritize sustenance fishing (12)	9
	Same as food fishing (13)		Use aquaculture (3)	
10	N/A	*Culture*	*Pono*/righteousness (14)	8
			Non-native practice (10)	

in Table 3 represent the number of occurrences of the subthemes within the emergent themes. Certain testimonies covered multiple themes and subthemes; therefore, the numbers in Table 3 do not equate to the number of individual testimonies.

Subthemes

Under the emergent theme *sustainability*, 'sustainable' and 'scientific support' were identified as subthemes by the opposition. 'Sustainable' indicated that the aquarium industry in its current form was sustainable. Testimony coded as 'sustainable' included comments such as, '[the aquarium trade] has proven to be sustainable at current levels' and 'we all want a vibrant, sustainable and diverse reef.' 'Scientific support' referred mainly to literature produced or referenced by the DLNR and/or its affiliates. Specifically, template testimony referred to a high-in-demand species, the yellow tang (*Zebrasoma flavenscens*), stating, 'its abundance over the entire West Hawaii coast has increased by over 1.3 million fish (up from 58%) from 1999 to 2013.' DLNR is a government organization tasked with the conservation and management of Hawai'i's natural resources. *Sustainability* subthemes for the supporters included the subthemes 'unsustainable' and 'noticeable decline in fish.' An example excerpt from template testimony coded as 'unstainable' was 'restores depleted populations of species heavily targeted by the aquarium trade.' The subtheme regarding 'noticeable decline in fish' referenced testimony with comments such as '... the fish population has declined dramatically from when we first came in the mid 90s' and 'I have heard many tourists say that Hawaii is no longer a viable place to come for scuba dive vacations [due to a perceived decline in fish].' There were six cases of 'reformed aquarium collectors' giving testimony on the legislation (Hawaii State Legislature , 2015); these individuals used personal experiences from inside the industry to emphasize their perceived unsustainability of the trade. One stated:

> I have seen the decline with my own eyes first hand. What was once a pristine secret reef - is now a barren waste land. What is a prime tourist dive site - is only left with the animals that are of little or no value to collectors.

'Become a model state' was in reference to Hawai'i's potential to follow suit of other nations in which the aquarium trade has been banned. Other subthemes indicated that some individuals felt the industry's sustainability could not be asserted without a standard 'Environmental Impact Statement' (EIS). Some felt additional 'Marine Protected Areas' (MPAs) would contribute to sustainability and a few indicated that 'more research was necessary.'

'Sufficient management' emerged as a subtheme under *regulations*. This indicated the opposition felt management and regulations were effective. For example, 'Hawaii's aquarium trade is considered to be one of, if not the best regulated nearshore fisheries in the state.' The supporters disagreed with the subtheme 'ineffective management' expressing, 'Current regulations have so far proven ineffective in protecting our coral reefs and surrounding habitats' and 'protect Hawaii from corporate abuse once again.' A subtheme unique to the opposition was 'leave it to the experts.' In general, this subtheme encompassed the idea that those working in management (e.g. DLNR) should be allowed to do their job without interference with comments such as 'Leave the regulation of this fishery to the experts and the DLNR!' However, another type of sentiment expressed under this subtheme was that 'Hawaiian aquarium fishermen are the experts and should be the ones certifying.'

Under *legislation*, a significant part of the opposition stated comments coded under 'poorly worded bill' with comments such as 'much of what it proposes is poorly defined and impossible to comply with.' Related to poor wording, was a separate subtheme 'coral not harvested.' HB 873 stated that coral extraction was occurring and that the trade was impacting Hawai'i's natural resources; thus, one version of opposition template testimony responded with 'There is no harvesting of coral in the state of Hawaii – it is forbidden and illegal to harvest or ship live rock of any kind. Nobody is doing this.' Another subtheme that emerged from both sides was 'confrontational language' describing statements that were confrontational towards the council. Excerpt examples from opposition were 'I can't believe you guys actually consider bills like this - so badly conceived, badly researched and badly done. I expect more from you all' and excerpts from the supporters, 'C'mon... wake up please.'

The emergent theme *economy* showed those in support of HB 873 perceived the aquarium trade as having a negative impact on the tourism and visitor industry (e.g. 'By preserving your reefs, you are preserving one of Hawaii's greatest assets, tourism'); such statements were coded under 'value of tourism.' Additionally, many self-identified tourists and divers expressing support of the legislation offered direct 'diver/tourist commentary,' as in that they want to see fish on the reef. Example excerpts include 'As a frequent visitor to the islands the highlight of our trips is snorkeling, viewing a different world and interacting with the marine life' and 'I am a diver that enjoys diving on the Big Island. I want to see these beautiful fish in their native environment, not in an aquarium.' Those in opposition of the legislation felt that further regulating the aquarium trade would have a 'negative impact on livelihoods and businesses.' Sentiments included statements, such as 'I strongly oppose any effort to further thwart the local economy.' An additional economic sentiment from the opposition was that some felt the aquarium trade 'supports tourism' by bringing aquarium collectors to the state so that they can see their captive fish in the wild on the reef.

Subthemes categorized under *user conflict* revealed that opposition felt further regulation on the aquarium industry meant the government was 'choosing an industry,' more specifically tourism. 'They are seeking to use the legislature to gain a business advantage, not protect Hawaii.' The subtheme 'emotional response' encompassed ideas that the supporters were crazy environmentalists or even radical extremists. Statements included, 'Please do not listen to the extremists' and 'Don't listen to the enviros.' Others opposing HB 873 stated feelings of love and enjoyment from participating in the aquarium trade or collecting marine organisms. Example excerpts included '[aquaria provide] thousands of people around the world an opportunity to enjoy the ocean and its creatures' and 'Why should the study and pleasure of fishing and fisheries and a aquariums be taken away in a free country?' Such sentiments were coded as 'personal enjoyment from the hobby.' With only a few mentions, some individuals from the opposition attempted to minimize resource-user conflict by stating that marine organisms are taken from reefs not frequented by tourists ('collection outside tourists' view). Alternatively, the supporters saw those involved in the aquarium industry and those opposing HB 873 as self-serving; such data was coded under 'greed.' Example comments were 'Living creatures should not be stolen to satisfy the greed of humans' and 'GREEDY aquarium takers do NOT GIVE BACK.' Another subtheme under user conflict described statements from supporters of HB 873 who were against the idea of using 'fish as ornaments.' Statements from supporters indicating 'violence' as an issue, referred to an incident in which a diver filming aquarium collectors was violently attacked underwater (see Mauinow.com, 2014).

'Concern for reef health' emerged as a subtheme from opposition under *social benefit*. This subtheme was influenced by template testimony that indicated aquarium collectors and those involved in the industry had a genuine concern for the reef health on which their hobby and/or employment depended, for example, 'We all want a vibrant, sustainable and diverse reef.' Other individuals suggested the dependence on the resource warranted the conservation of it with example excerpts such as 'No group is more invested in a healthy reef than fishermen that depend on the ocean for their living.' This subtheme was coded differently than sustainability due to the personal concern expressed for the reef. Some opposing HB 873 felt the aquarium industry provided educational benefits through observation, or that aquariums are our last hope for conservation ('Education/last hope'). For example, one person stated: 'aquariums are a great way to show people how beautiful and special fish are' while another felt that due to the existing destruction of the environment, 'aquariums and captive breeding is the last resource to maintaining all these beautiful species.' The idea of 'collective resources' was a subtheme under both supporters ($n = 268$) and the opposition ($n = 2$). This subtheme referred to the idea that marine resources were a collective resource (e.g. supporter's template excerpt: 'our marine wildlife'). Many felt preserving the resources for 'future generations' was critical (e.g. template excerpt coded as both collective resources and future generations: 'protect our reefs for generations to come'). Statements from supporters of HB 873 addressing ethics of the aquarium trade were coded as 'unethical.' Examples included a recurring labeling of the industry as a crime against nature and statements, such as 'It is a terrible sin to capture a live animal and make it suffer in a small aquarium, often to die quickly and painfully after leaving the beautiful reef where it was "stolen"' and 'It broke my heart to see this beautiful Hawaiian reef fish [Humuhumunukunukuapua'a] in such a small, artificial habitat in a grey, gravelly, armpit of a town thousands of miles away from home.' There were two cases from the support that indicated support for the aquarium trade only for purposes of education through public aquaria ('Public aquaria only').

The idea of other environmental issues outside of the aquarium trade that negatively affect the marine environment was present in both the opposition and the supporters. The opposition felt that 'other environmental issues are more pressing' with a member from the opposition stating:

> The egregious nature of the allegations presented in this bill only illuminates more clearly the attempt to lay any impact to Hawaii's reefs at the feet of ignoring the true stressors to the coastal marine environment, of near shore development, runoff, injection wells, and heavy tourism.

Alternatively, the supporters recognized that many 'other environmental issues' do affect the marine environment making it necessary to take any measure possible to conserve the resource noting that the reefs are 'suffering already from [global] warming and toxic runoff.' Additionally, the supporters used statements suggesting the marine environment is a 'complex ecosystem,' to call for an increased urgency of creating additional environmental protections. For example, 'The oceanic ecosystem is a complex system not fully understood.'

Under the theme *Animal welfare*, the opposition expressed using the best possible techniques and husbandry or 'best practices' in capturing marine organisms stating 'Great steps have been taken in shipping and husbandry of aquarium inhabitants.' From the supporters, the template testimonies, as well as individuals, addressed ideas related to the 'high mortality' (e.g. template testimony excerpt: 'Please amend to ensure that it greatly improves the treatment and respect for our marine wildlife' in the industry and 'inhumane' treatment of the animals (e.g. puncturing swim bladders, fin clipping, starvation) with excerpts such as 'I have seen fish put into tiny boxes mean for organizing of screws and hardware.' Some chose to cite animal welfare laws such as a statement: 'Every state in the Union prohibits endangering or physical abuse of wildlife.'

Within the theme *extractive resource use*, data from the opposition demonstrated that some felt that the aquarium industry was a fishing sector, hence the idea that it cannot or should not be banned ('don't ban fishing'). Similarly, others felt that the context of resource extraction was irrelevant and that removal of marine organisms for the aquarium industry was the 'same as food fishing' with testifiers noting 'Anti-trade activists turn a blind eye to the harvest of a species like the Achilles Tang for food, but cry foul if the same fish were to be harvested for the aquarium trade.' Those in support, felt it important to 'prioritize sustenance fishing,' explaining 'reef ecosystems also provide sustenance for many of our families in Hawai'i.' Such comments suggest that supporters view the aquarium industry as different from a food fishery. Alternatively, some supporters felt the aquarium industry should turn to or 'use aquaculture.'

The theme *culture* was only mentioned by those supporting HB 873. It is further noted that some members of the support self-identified as native Hawaiian. Data were coded as either 'pono' or 'not a native practice.' *Pono* (righteousness) is a cultural idea and practice. This was communicated with statements, such as 'Wasting wildlife is not *pono*.' Data coded as non-native practice included straightforward statements (e.g. it's not a native practice) and more elaborate references to traditional Hawaiian knowledge (as well as the idea of prioritizing sustenance fishing), 'The KAPU system [traditional Hawaiian resource management] was for protecting food stocks–here with the Aquarium Fish Collection Trade we are not talking about people fishing to feed their families and/or the community.' Other statements coded under 'non-native practice' described potential cultural damage from the taking of an *'aumakua* by a collector. For example, 'These fish and other sea animals that are attractive to the [aquarium trade] trade such as *Puhi* (eels) are being taken – many of these animals are *'Aumakua* belonging to different families.' *'Aumakua*, simplified, is the idea of a guardian spirit in the form of an animal. It refers to a single and specific animal, rather than an entire species. Marine species as *'aumakua* are common.

Rank

For comparative purposes, a rank was assigned to themes for opposition and supporters indicating the number of occurrences within the entire data-set (Table 3). It is noted that the opposition, when compared to the supporters, relied more heavily upon template testimony (refer to Figure 1). The top three ranking themes from the opposition were significantly influenced by template testimony with template testimony accounting for the following percentages of the top three opposition themes: sustainability (74%), regulations (84%), and legislation (90%). The top three themes under the supporters group were also influenced by template testimony, though to a lesser extent, with template comments accounting for the following of the top three themes: social benefit (34%), animal welfare (81%), and sustainability (51%). For both opposition and supporters, *economy* and *user conflict* were ranked as fourth and fifth, respectively. Opposition testimonies produced *social benefit* as rank 6, while supporters placed *environment* as rank 6. *Environment* ranked seventh from the opposition and *regulations* ranked seventh from the supporters. The eighth ranking was assigned to *animal welfare* for the opposition and *culture* for the supporters. *Extractive resource use* was ranked ninth from both parties. The opposition did not have a 10th rank as none of the testimonies from this data-set mentioned culture. For the supporters, *legislation* was assigned a 10th rank.

Discussion

The data demonstrated that both sides felt strongly about their respective positions. There were many notable areas of thematic overlap within the subthemes and additionally, commonalities with importance ranking of the themes. The expressed commonalities, as identified through thematic overlap, may be useful for negotiating or mediating future conservation efforts between marine resource-user groups (Wall, Stark, & Standifer, 2001). As this was a qualitative analysis, it is difficult to quantify the perceived importance of themes beyond an assigned rudimentary rank. This is further impacted by the influence of template testimonies on the occurrence and frequency of themes and subthemes (refer to Figure 1). Template testimony from the opposition accounted for over half of theme occurrences of the following themes: (a) *sustainability*; (b) *regulations*; (c) *legislation*; (d) *economy*; (e) *user conflict*, (f) *social benefit*, and (g) *environment*. Alternatively, template testimonies from the supporters accounted for over half of theme occurrences of only two themes: (a) *animal welfare* and (b) *sustainability*, and a notable part of *social benefit*. NIMBY-type constructs were evident within certain themes. For example, organizational distrust was evident under *legislation*, access to information in *sustainability*, risk aversion and localization of the issues in both *economy* and *environment*, and emotional attachments under *user conflict*. Each emergent theme is discussed individually.

Sustainability

Sustainability appeared a key issue for both the opposition and the supporters; yet, for different reasons. In general, the opposition viewed the aquarium trade as sustainable. 'Science,' as perceived by the opposition refers to findings from specific studies out of Hawai'i (e.g. Tissot et al., 2009; Walsh, 2015; Walsh et al., 2003). Indeed, these studies reported positive results for *some* measured species (e.g. increase in yellow tang populations) from FRAs or ecosystem-based management, yet, all of the referenced studies disclosed limitations of the current science and/or regulations and enforcement. Some supporters questioned the validity of these studies due to at least one author/investigator being involved in marine resource management at the Hawai'i state level. Multiple other studies, including those used to support the oppositions' positions, reveal that the data on the subject are limited or incomplete, making it challenging to ascertain actual sustainability of the trade (Albins, & Hixon, 2013; Rhyne et al., 2012; Kolm & Berglund, 2003; Tissot & Hallacher, 2003; Tissot et al., 2009; Walsh, 2015; Walsh et al., 2003). Even if future research adequately addresses these gaps, using science for policy is not straightforward. Fischer (2000) describes the challenge in using science for policy-making, stating issues of precision and also influences of different disciplines, thus complicating the ability of science to adequately address many environmental questions.

Alternatively, many supporters described the trade as non-sustainable, citing a noticeable decline in fish. While the accounts of fish decline are anecdotal and based on personal observation, they may be worthy of further attention. Local knowledge can be valuable (Fischer, 2000; Priskin, 2003), yet citizen science is often challenged by organizational issues and issues relating to data collection and use (Ottinger, 2010).

Regulations

Regulations was another theme heavily influenced by template testimony on the side of the opposition. In particular, the general perception that the aquarium industry was subject to 'sufficient management' and 'leave it to the experts.' Fischer (2000) suggested: 'Those who have problems with participatory solutions typically argue that the elite professionals should govern in the interest of competence... For many, the public emerges as something to worry about, if not fear' (p. 33). Being that this theme did not appear under the supporters, whom in a few cases demanded more science (see *Sustainability* theme), it is likely that the idea of deferring to experts is subjective and based on current agreement of the 'experts' and opposition. Subthemes from supporters included 'ineffective management' which translated to ideas such as under reporting and corruption. Under reporting has been referenced in the literature (see Walsh et al., 2003), though more recent works have reported an increase in accuracy in reporting (see Rhyne et al., 2012; Walsh, 2015).

Legislation

Many from the opposition were dissatisfied with the wording of and/or statements in HB 873. This was interpreted as result of perceived falsities in the legislation referencing the absence of coral harvesting. The harvesting of live rock and corals is prohibited under Hawai'i law. There is little evidence to support the occurrence of a live coral trade originating in Hawai'i. Another notable difference was the occurrence of confrontational language. The confrontational language toward the council was more prevalent from the opposition ($n = 77$) compared to the supporters ($n = 3$). Such language is described by Croom (2013) as expressive. The difference in linguistic patterns suggests a diversity of social/community groups between the opposition and supporters and possible social stratification (see Croom, 2013, for in-depth review).

Economy

The dissension between the opposition and supporters on economics is interesting when viewed as a factor of economic scale. This section is devoted to a discussion on the main two competing sectors, the aquarium trade and tourism. For the year 2015, DLNR-DAR issued a total of 356 non-commercial permits and 200 commercial aquarium permits; of the commercially licensed collectors only 175 were 'active' (e.g. regularly reporting catches) and thus, required to submit commercial aquarium fish reports (R. Kokubun, personal communication, 29 November 2016).

Comparably, the tourism industry employs over 170,000 statewide (HTA, 2016). Monetarily, tourism sales are estimated annually at $25.2 billion (HTA, 2016), while the aquarium industry has been valued at $3.2 million (Hawaii Coral Reef Initiative, n.d.). Further, Rossiter and Levine (2014) suggest that economic benefits from the aquarium trade are 'only realized by a very small (and often non-indigenous) sub-set of the local population' (p. 199). There was some mention from the opposition that tourism industries benefit from the aquarium trade by drawing visitors interested in seeing natural habitats of their captive fish, although occurrence of testimony from supporters suggesting the importance of a non-harvested reef was far greater. Snorkeling and diving are important recreation activities for the state, with nearly 44% of tourists participating in these activities (from HTA; 2015). Interestingly, it seems that many tourists/visitors expressed vested interests in the state's marine resources testifying that they wanted to see fish on the reef, rather than in a tank. While for those involved in the industry, the idea of increased regulations or bans threatens their immediate livelihoods, the testimonies originating from reformed aquarium collectors ($n = 6$) (see *Sustainability* theme) suggest the potential for successful livelihood shifts. Though the number of livelihoods affected by regulations would be minimal,

Stevenson, Tissot, and Walsh (2013) showed that overall socioeconomic well-being consequences associated with previous regulations imposed on the Hawai'i industry (similarly to Christie, 2004), were not detectable. They also suggested a consideration of reducing fishing efforts, though noted the complexity in altering livelihoods stating, 'this may not always be politically or socially feasible, but it may be appropriate and necessary in some cases' (p. 57).

User conflict

Subthemes that emerged under user conflict were notably emotional. Some of the opposition viewed the supporters as extremists, while some supporters viewed the opposition as greedy. Radical behavior is not new to activism (Fischer, 2000); however, the opposition did not provide examples to support these labels within the testimony and no such examples were found in the media. The view that the opposition was greedy was founded in the personal gain (financial or ornamental) of individuals or limited audiences profiting from the removal of the 'collective' marine resources for aquarium purposes (reported in Rossiter & Levine, 2014). This thinking is in line with tourism literature that suggests the potential conservation benefits of tourism development through non-extractive resource use (see Samonte-Tan et al., 2007). A recurrent idea in the individual testimonies of the opposition was showing preference to the tourism sector. This sentiment highlights the direct user conflict between the aquarium and tourism and recreation activities. Due to rapidly changing natural environments, prioritizing an industry (tourism, aquarium trade, or otherwise) will likely become necessary in the near future (see Stevenson et al., 2013).

Social benefit and environment

For discussion purposes, social benefit and environment have been combined as they are interdependent. Social benefit was the most mentioned theme by the supporters who repeatedly stated an obligation to protect the marine resources for the greater good. This sentiment mirrors that of a large NOAA study that demonstrated U.S. residents were concerned for and willing to pay to conserve Hawai'i reefs (Bishop et al., 2011). Contrastingly, the opposition viewed the aquarium trade and its effects on a much smaller and personal scale. For example, the opposition repeatedly encouraged the council to focus on other environmental issues, whereas the supporters recognized other environmental issues and noted the importance of doing everything possible, including managing the aquarium trade, to conserve the marine resources. Fishing, marine resource extraction, tourism, and recreational uses of the marine environment all impact reef ecosystems. A study from Hawai'i demonstrated a negative correlation between fish biomass and local human population density where shorelines were easily accessible (Williams et al., 2008). Issues associated with tourism misuse or overuse such as reef trampling, effects of human effluent, and walling of shorelines have been well studied in the literature making their impacts to the reef environment more easily measured (e.g. Bulleri & Chapman, 2010; Hannak et al., 2011; van Beukering & Cesar, 2004; Williamson et al., 2017). Whereas, the impacts associated with the removal of marine species for the aquarium trade remain understudied making it impossible to determine the impacts or sustainability for most of the targeted populations (Rhyne et al., 2012).

This area of dissension requires further attention. From an ecological vantage point, understanding the targeted populations is critical. From a social vantage point, it is necessary to identify not only how stakeholders are identifying with the marine resources, but the underlying justifications for continued resource use and/or extraction. The literature on 'last chance tourism' may offer some insight toward the opposition. The idea of 'get it before its gone' is a driver in some tourism locations. For example, Piggott-McKellar and McNamara (2016) found nearly 70% of surveyed tourists visiting the Great Barrier Reef were motivated by last chance tourism; however, very few understood the impacts of the personal behaviors (e.g. effects of tourism/travel to the destination). More research is needed to understand the accuracy of the aquarium collectors and hobbyists' perceptions of personal impacts resulting from the trade or if the 'last chance' construct is a motivating factor for aquarium collectors. It is, without doubt, known that healthy reef ecosystems would benefit all marine resource users. However, due to the complexity of the marine ecosystems, the mobility of fish, and other anthropogenic variables (e.g. underreporting, runoff, climate change), limiting the necessary variables to understand the specific impact of the aquarium trade will continue to be difficult.

Animal welfare

Mortality in the aquarium trade can be high. A study from Indonesia revealed post-harvest mortality rates of 10%–40% (Schmidt & Kunzmann, 2005). It is in the best interest of those involved in the trade to decrease mortality. Practices such as clipping fin tips (to avoid puncturing bags during shipping) and starving organisms for a day or more prior to shipping (to maintain water quality during transit) are part of the husbandry practiced by the trade. Other known practices include venting (puncturing swim bladders to rapidly decompress fish) are used at the time of capture to reduce decompression time of the organisms while surfacing from the dives (see Munday, Tissot, Heidel, & Miller-Morgan, 2014). Munday et al. (2014) found that venting did not increase mortality; however, they did note a stress response increase in vented fishes. The opposition perceived current husbandry practices as the best practices available. Alternatively, supporters viewed these practices as animal cruelty. Venting and fin clipping are not addressed under state animal cruelty laws; however, starvation is defined as animal cruelty for all animals other than insects, vermin, and other pests (HI Rev Stat § 711-1109, 2013). Many agree that fish are capable of feeling or that sentience in fish cannot be adequately disproven (e.g. Broom, 2017; Brown, 2016; Damasio & Damasio, 2016; Sneddon & Leach, 2016). The elevated cortisone levels (stress response) found in Munday et al. (2014) following venting may further support this. The recent research on fish sentience, coupled with state law, validates the supporters' concerns for animal welfare.

Extractive resource use

Some of the opposition perceived the aquarium trade as a fishery, indicated by statements requesting to not ban fishing. Alternatively, some supporters were against classifying the aquarium trade as a fishery associating aquarium collection with entertainment rather than sustenance. The term fishery was originally associated with extractive sustenance and subsistence use and has evolved to encompass other types of extractive marine resource use including live fish trades (e.g. live fish reef trade (LFRT)). Currently, the literature refers to the aquarium trade as both a fishery (Walsh et al., 2003) and a trade (Rhyne et al., 2012), and in some cases both (Kolm & Berglund, 2002). Given that the goal of the aquarium industry is to collect live specimens, it is seemingly most similar to the LFRT, which is both a fishery (for statistical purposes) and a trade (for economic purposes). Both the LFRT and food fisheries end with sustenance, while the aquarium trade ends in observation. Thus, the aquarium trade is discernibly different from a food fishery, even the LFRT. Due to the economic exchanges associated with the industry, labeling it as an aquarium trade fishery should be standard, but perhaps even more appropriate would be naming it an observation fishery or observation trade fishery since the end result is some form of observation, rather than consumption. Identifying aquarium collection as a practice distinguishable from other fisheries may provide insightful impacts beyond the environment, such as culture.

Culture

Hawai'i is unique, not only geographically, but also culturally. Hawai'i is known for its 'Aloha' spirit (Mak, 2015). Despite the well-described linkages between fishing and culture in Hawai'i (Goto, 1986), culture emerged only from supporters' comments. The concept of *pono*, or righteousness, is commonly discussed and applied in Hawai'i (see Tissot, 2005) and has been used at a community level to enact management practices for the marine environment in the absence of formal rules (Friedlander, Shackeroff, & Kittinger, 2013). In pre-contact Hawai'i, island societies developed and successfully followed complex and effective systems for marine resources management (see Friedlander et al., 2013). As stated by the supporters, of whom some identified as Native Hawaiian Practitioners, aquarium collection is not a native Hawaiian practice. Culture was not mentioned by the opposition. It is unclear if this element was simply overlooked or purposely avoided. Though the data from this theme is limited to the supporters, the description of Tissot (2005), regarding a 'classic conflict between the Judeo-Christian worldview that resources are for our use (i.e. the Garden of Eden) and the Hawaiian philosophy of *mālama 'āina* (caring for the land)' is applicable (p. 89). More research is necessary to understand how, or if, the opposition views the aquarium trade with regard to Hawaiian culture.

Thematic summary

The interconnectedness of the themes was apparent with areas of overlap between themes. Many of emergent themes were directly entwined with tourism (e.g. extractive resource use, user conflict, sustainability, economy, culture, social benefit and environment). Additionally, the spatial demographics of the supporters demonstrated the closeness of tourism to the aquarium collection issue. These findings suggest that tourists feel responsibility towards destination conservation and a willingness to participate in environmental issues. Their interest in the aquarium issue and the general marine resources of Hawai'i is likely a result of having previously visited Hawai'i. Globally, numerous individuals have identified themselves as stakeholders in the marine resources of Hawai'i. This is in contrast to the opposition which was a largely localized [to Hawai'i] stakeholder group, despite global participation in the aquarium hobby. In general conflicts between tourism and the local community are expected (Hannak et al., 2011); however, the data from this study revealed that sections of the local community shared sentiments similar to tourists (see also Big Island Now.Com, 2017; Muller, 2012). An ethical conflict surrounding the aquarium trade was also revealed in the data. Many supporters (including tourist stake-holders) struggled to justify the taking of individual marine organisms for personal profit or enjoyment. This indicates that efforts future efforts towards resource-use management should consider the issue an ethical conflict, as well as an environmental issue.

Limitations

Research using public testimony is still an evolving area. Thus, there are important limita-tions associated with this piece of research. The first and foremost, this research was lim-ited by the data-set. While public testimony offers an exploratory vantage into issues through unsolicited (by the researcher) data (Fiorinio, 1990), it does not account for all resource-user groups, nor is it representative of the general population. Further, the broader themes and subthemes were derived from hundreds of individual testimonies; therefore, these themes and subthemes offer generalizations of the groups and may not be entirely representative at an individual level. Additionally, it was impossible to deter-mine the authenticity of individual comments. This concern was twofold. First, given the format of the data as online records, it was impossible to assess the trustworthiness of individuals' comments as a means of reflecting true inner feelings and thoughts. Second, given the nature of the outlets for submission (e.g. online), it is possible that individual records may have been falsified. Although public testimony on HB 873 was open to any individual, it is generally only those close to the issue who make the effort to submit testi-mony. There were no official advertising efforts associated with HB 873; therefore, only those subscribed to a listserv/organization with a vested interest in the issue were notified or those individuals who researched the bill/issue independently.

Another limitation of the research was the researcher's closeness to the issue. The researcher openly communicates a personal ethic against the 'need' for ornamental aquar-ium collection. The researcher's personal testimony in support of HB 873 (a combination of individual comments and template testimony) was included in the data-set. It is also noted that the researcher is in acquaintance with some of the persons who provided testimonies. Personal relations were not considered to have influenced the data analysis as names were removed from the data-set following the removal of duplicate testimonies and prior to cod-ing of the data. The researcher's closeness to the issue of ornamental aquarium collection provided a motive for rigor within the thematic analysis and in the identification of themes, likely more so than had the researcher had a neutral opinion on the issue.

Conclusions

Conflict resolution requires cooperation (Stepanova, 2015). This paper has highlighted the perceived areas of concern from user conflict between tourists, residents, recreationists, aquarium collectors, aquarium dealers, and hobbyists resulting from the aquarium trade in Hawai'i. Though conflicts arising from resource use are most commonly addressed with public consultation (e.g. the process used in HB 873) (Stepanova, 2015), new resource management strategies and approaches are needed in Hawai'i as the current, Western-based management schemes do not mesh well with the socio-ecological contexts of island state (Friedlander et al., 2013). Undertaking a thematic analysis of public testimony is time consuming; yet, it can serve to transform comment from opinion to applied strategy. The findings from this research suggest that various resource-user group opinions can be supported or addressed through targeted future research. These are summarized here individually: (1) the arguments regarding sustainability of the trade can be settled by future, unaffiliated (with state marine management) research efforts; (2) the validity of arguments citing a decline in fish can be addressed with qualitative research that explores tourist and recreationist perceptions on fish populations without reference to any particular environmental impact (e.g. the aquarium trade); (3) the potential social stratifications should be investigated through demographic research as understanding these will likely become relevant for future community-based resource management schemes; (4) the debate about 'which sector is worse' can be addressed using technology such as HD underwater cameras that monitor high traffic areas for both the tourism and aquarium trade; and (5) given the cultural and contextual confusion surrounding the practice of extracting marine organisms for aquarium purposes, this practice should be more descriptively labeled (e.g. live fish trade). Additionally, the basic areas of thematic overlap may serve as foundations for future discussions between the resource-user groups. For example, the concern with reef health is an area of strong agreement and may be a beneficial starting point for future attempts to advance the discussion regarding resource-user conflict. It is plausible that both the support and opposition of HB 873 may be supportive of establishing population assessments for targeted species.

Transforming public comments into useable knowledge clusters, or themes, creates a foundational knowledge integration for resolution strategy (Stepanova, 2015). While research from the natural sciences regarding reef populations remains necessary, there is a parallel need for more qualitative data on the issue of the aquarium trade. To date, one of the largest stakeholder groups, tourists, have been largely excluded from this debate. The findings from this research suggest that perception-based data from tourist stakeholders may become a critical social component to future policy surrounding this issue (see Bishop et al., 2011). This research has shown that Hawai'i tourists self-identify as stakeholders in the marine resources of Hawai'i. Specifically, those without a vested interest in the aquarium trade showed interest in increased protection for the state's marine resources. If Hawai'i is to remain the 'Paradise of the Pacific' (see Mak, 2015), a better understanding of tourist perceptions on the aquarium trade industry is needed.

Acknowledgement

First and foremost, the author would like to thank all of the persons who spoke out on this issue. Their testimonies provided the data set for this research. The author would also like to thank the reviewers for their comments and inputs which helped to improve and clarify this manuscript.

Disclosure statement

No potential conflict of interest was reported by the author.

ORCID

Brooke A. Porter ⓘ http://orcid.org/0000-0002-8183-9662

References

Albins, M. A., & Hixon, M. A. (2013). Worst case scenario: Potential long-term effects of invasive preda-tory lionfish (*Pterois volitans*) on Atlantic and Caribbean coral-reef communities. *Environmental Biology of Fishes, 96*(10–11), 1151–1157.

Basit, T. (2010). Manual or electronic? The role of coding in qualitative data analysis. *Educational Research, 45*(2), 143–154.

Bell, J. (2005). *Doing your research project: A guide for first-time researchers in education, health and social science.* Berkshire, England: Open University Press.

Big Island Now.Com (2017, June 6). *90 percent of residents favor reef protections from aquarium trade.* . Retrieved from http://bigislandnow.com/2017/06/06/90-percent-of-residents-favor-reef-protec tions-from-aquarium-trade/

Bishop, R. C., Chapman, D. J., Kanninen, B. J., Krosnick, J. A., Leeworthy, B., & Meade, N. F. (2011). *Total economic value for protecting and restoring Hawaiian coral reef ecosystems : Final report* (NOAA Technical Memorandum CRCP 16). Silver Spring, MD: NOAA Office of National Marine Sanctuaries, Office of Response and Restoration, and Coral Reef Conservation Program.

Braun, V., & Clarke, V. (2006). Using thematic analysis in Psychology. *Qualitative Researchin Psychol-ogy, 3*(2), 77–101.

Broom, D. M. (2017). Fish brains and behaviour indicate capacity for feeling pain [Peer commentary on "Why fish do not feel pain" by B. Key]. *Animal Sentience: An Interdisciplinary Journal on Animal Feeling, 3*(1), 2017.017

Brown, C. (2016). Fish pain: An inconvenient truth [Peer commentary on "Why fish do not feel pain" by B . Key]. *Animal Sentience: An Interdisciplinary Journal on Animal Feeling, 3*(1), 2016.058.

Bulleri, F., & Chapman, M. G. (2010). The introduction of coastal infrastructure as a driver of change in marine environments. *Journal of Applied Ecology, 47*(1), 26–35.

Christie, P. (2004). Marine protected areas as biological successes and social failures in Southeast Asia. In J. B. Shipley (Ed.), *American fisheries science symposium* (Vol. 42), pp. 155–164. Bethesda, MD: American Fisheries Society.

Croom, A. M. (2013). How to do things with slurs: Studies in the way of derogatory words. *Language & Communication, 33*(3), 177–204.

Damasio, A., & Damasio, H. (2016). Pain and other feelings in humans and animals [Peer commentary on "Why fish do not feel pain" by B. Key]. *Animal Sentience: An Interdisciplinary Journal on Animal Feeling, 3*(1), 2016.059.

Fabinyi, M. (2008). Dive tourism, fishing and marine protected areas in the Calamianes Islands, Philip-pines. *Marine Policy, 32*(6), 898–904.

Fereday, J., & Muir-Cochrane, E. (2006). Demonstrating rigor using thematic analysis: A hybrid approach of inductive and deductive coding and theme developmentl. *International Journal of Qualitative Methods, 5*(1), 80–92.

Fiorino, D. J. (1990). Citizen participation and environmental risk: A survey of institutional mecha-nisms. *Science, Technology, & Human Values, 15*(2), 226–243.

Fischer, F. (2000). *Citizens, experts, and the environment: The politics of local knowledge.* Durham, NC: Duke University Press.

Friedlander, A. M., Shackeroff, J. M., & Kittinger, J. N. (2013). Customary marine resource knowledge and use in contemporary Hawai'i. *Pacific Science, 67*(3), 441–460.

Froehlich, H. E., Gentry, R. R., Rust, M. B., Grimm, D., & Halpern, B. S. (2017). Public perceptions of aquaculture: Evaluating spatiotemporal patterns of sentiment around the world. *PloS One, 12*(1), e0169281.

Goto, A. (1986). *Prehistoric ecology and economy of fishing in Hawaii: An ethnoarchaeological approach* (Unpublished doctoral dissertation). Honolulu, HI: University of Hawai'i.

HI Rev Stat § 711-1109. (2013). (*Cruelty to Animals in the Second Degree*, 37 H. R. S. §711-1109).

Hannak, J. S., Kompatscher, S., Stachowitsch, M., & Herler, J. (2011). Snorkelling and trampling in shallow-water fringing reefs: Risk assessment and proposed management strategy. *Journal of Environmental Management, 92*(10), 2723–2733.

Hawaii Coral Reef Initiative. (n.d). *Economic value of Hawaii's nearshore reefs.* Retrieved from http://www.hawaii.edu/ssri/cron/files/econ_brochure.pdf

Hawaii State Legislature. (2015). *H. 873, 28th Legislature.* Retrieved from http://www.capitol.hawaii.gov/Archives/measure_indiv_Archives.aspx?billtype=HB&billnumber=873&year=2015

Hawaii Tourism Authority. (2016). *Hawaii tourism fact sheet.* Retrieved from http://www.hawaiitourismauthority.org/default/assets/File/021016%20Fact%20Sheet_REV.pdf

Kolm, N, , Berglund, A. (2003). Wild populations of a reef fish suffer from the "nondestructive" aquarium trade fishery. *Conservation Biology, 17*(3), 910–914.

Kolm, N., & Berglund, A. (2002). Wild population of a reef fish suffer from the 'nondestructive' aquarium trade fishery. *Conservation Biology, 17*(3), 910–914.

Majanen, T. (2007). Resource use conflicts in Mabini and Tingloy, the Philippines. *Marine Policy, 31*(4), 480–487.

Mak, J. (2015). Creating "Paradise of the Pacific": How Tourism Began in Hawaii (Working Paper No. 15-03). Honolulu, HI: University of Hawai'i at Manoa, Department of Economic Working Paper Series.

Mauinow.com (2014, May 13). *Video: Diver attacked while filming reef fish collector. Maui Now* Retrieved from http://mauinow.com/2014/05/13/video-diver-attacked-while-filming-reef-fish-collector/

Miles, M. B., Huberman, A. M. (1994). *Qualitative Data Analysis,* 2nd. Thousand Oaks, CA: Sage.

Muller, E. (2012, June 14). Most residents want aquarium fish collecting banned, poll says. *Hawaii Tribune Herald.* Retrieved from http://hawaiitribune-herald.com/sections/news/local-news/most-residents-want-aquarium-fish-collecting-banned-poll-says.html

Munday, E. S., Tissot, B. N., Heidel, J. R., & Miller-Morgan, T. (2015). The effects of venting and decompression on Yellow Tang (*Zebrasoma flavescens*) in the ornamental aquarium fish trade. *PeerJ, 3: e756.* doi:10.7717/peerj.756

Nadon, M. O. (2017). *Stock assessment of the coral reef fishes of Hawaii, 2016* (NOAA Tech. Memo., NOAA-TM-NMFS-PIFSC-60). Honolulu, HI: U.S. Department of Commerce.

Ogawa, T., & Brown, C. L. (2001). Ornamental reef fish aquaculture and collection in Hawaii. In C. L. Brown & L., Young (Eds.), Marine Ornamentals '99: Proceedings of the First International Meeting; Kona, HI. Special Vol., Aquarium Sciences and Conservation 3(1–3), (pp 151–169). London: Kluwer Publications.

Ong, L. T. J., Storey, D., & Minnery, J. (2011). Beyond the beach: Balancing environmental and socio-cultural sustainability in Boracay, the Philippines. *Tourism Geographies, 13*(4), 549–569.

Ottinger, G. (2010). Buckets of resistance: Standards and the effectiveness of citizen science. *Science, Technology, & Human Values, 35*(2), 244–270.

Packer, J., & Ballantyne, R. (2002). Motivational factors and the visitor experience: A comparison of three sites. *Curator: The Museum Journal, 45,* 183–198.

Piggott-McKellar, A. E., & McNamara, K. E. (2017). Last chance tourism and the Great Barrier Reef. *Journal of Sustainable Tourism, 25*(3), 397–415.

Priskin, J. (2003). Tourist perceptions of degradation caused by coastal nature-based recreation. *Environmental Management, 32*(2), 189–204.

Rhyne, A. L., Tlusty, M. F., Schofield, P. J., Kaufman, L., Morris, J. A., Jr., & Bruckner, A. W. (2012). Revealing the appetite of the marine aquarium fish trade: The volume and biodiversity of fish imported into the United states. *PLoS One, 7*(5), e35808.

Rossiter, J. S., Levine, A. (2014). What makes a successful marine protected area? The unique context of Hawaii's fish replenishment areas. *Marine Policy, 44*(2014), 196–203.

Samonte-Tan, G. P. B., White, A. T., Tercero, M. A., Diviva, J., Tabara, E., & Caballes, C. (2007). Economic valuation of coastal and marine resources: Bohol Marine Triangle, Philippines, *Coastal Management, 35*(2), 319–338.

Schmidt, C., & Kunzmann, A. (2005). Post-harvest mortality in the marine aquarium trade: A case study of an Indonesian export facility. *SPC Live Reef Fish Information Bulletin, 13*, 3–12.

Sneddon, L. U., & Leach, M. C. (2016). Anthropomorphic denial of fish pain [Peer commentary on "Why fish do not feel pain" by B. Key]. *Animal Sentience: An Interdisciplinary Journal on Animal Feeling, 3*(1), 2016:035.

Spicker, P. (2007). Research without consent. *Social Research Update, 51*. Retrieved from http://www.soc.surrey.ac.uk/sru/.

Stepanova, O. (2015). Conflict resolution in coastal resource management: Comparative analysis of case studies from four European countries. *Ocean & Coastal Management, 103*, 109–122.

Stevenson, T. C., Tissot, B. N., & Walsh, W. J. (2013). Socioeconomic consequence of fishing displacement from marine protected areas in Hawaii. *Biological Conservation, 160*, 50–58.

Tisson, B. N., & Hallacher, L. E. (2003). Effects of aquarium collectors on coral reef fishes in Kona, Hawaii. *Conservation Biology, 17*(6), 1759–1768.

Tissot, B. N. (2005). Integral marine ecology: Community-based fishery management in Hawai'i. *World Futures, 61*(1–2), 79–95.

Tissot, B. N., Walsh, W. J., & Hixon, M. A. (2009). Hawaiian Islands marine ecosystem case study: Ecosystem-and community-based management in Hawaii. *Coastal Management, 37*(3–4), 255–273.

Voyer, M., Barclay, K., McIlgorm, A., & Mazur, N. (2017). Connections or conflict? A social and economic analysis of the interconnections between the professional fishing industry, recreational fishing and marine tourism in coastal communities in NSW, Australia. *Marine Policy, 76*, 114–121.

Wall, J. A. Jr., & Stark, J. B., & Standifer, R. L. (2001). Mediation: A current review and theory development. *Journal of Conflict Resolution, 45*(3), 370–391.

Walsh, W. J. (2000). *Aquarium collecting in West Hawaii: A historical overview*. Honolulu, HI: Department of Land and Natural Resources, Division of Aquatic Resources, Island of Hawaii. Retrieved from http://www.coralreefnetwork.com/kona/Walsh%20Aquarium%20Overview%202000.pdf

Walsh, W. J. (2015). *Report to the Thirtieth Legislature 2015 Regular Session: Report on the findings and recommendations of effectiveness of the West Hawai'i Regional Fishery Management Area* (Report). Retrieved from http://dlnr.hawaii.gov/dar/files/2015/01/ar_hrs188_2015.pdf

Walsh, W. J., Cotton, S. S., Dierking, J., & Williams, I. D. (2004). The commercial marine aquarium fishery in Hawai'i. In *Status of Hawai'i's coastal fisheries in the new millennium*. Proceedings of the 2001 fisheries symposium sponsored by the American Fisheries Society, Hawai'i Chapter, 2004 revised edition, A. M. Friedlander (pp. 129–156, Honolulu, HI: Hawaii Audubon Society).

Welsch, E., (2002). Dealing with data: Using NVivo in the qualitative data analysis process. *Qualitative Social Research, 3*(2), Art. 26.

Williams, I. D., Walsh, W. J., Schroeder, R. E., Friedlander, A. M., Richards, B. L., & Stamoulis, K. A. (2008). Assessing the importance of fishing impacts on Hawaiian coral reef fish assemblages along regional-scale human populations gradients. *Environmental Conservation, 35*(3), 261–272.

Williamson, J. E., Byrnes, E. E., Clark, J. A., Connolly, D. M., Schiller, S. E., Thompson, J. A., ... Raoult, V. (2017). Ecological impacts and management implications of reef walking on a tropical reef flat community. *Marine Pollution Bulletin, 114*(2), 742–750.

Wolsink, M. (2006). Invalid theory impedes our understanding: A critique on the persistence of the language of NIMBY. *Transactions of the Institute of British Geographers, 31*(1), 85–91.

van Beukering, P. J. H., & Cesar, H. S. J. (2004). Ecological economic modeling of coral reefs: Evaluating tourist overuse at Hanauma Bay and algae blooms at the Kīhei Coast, Hawai'i. *Pacific Science, 58*(2), 243–260.

Index

Note: Page numbers in *italics* refer to figures
Page numbers in **bold** refer to tables
Page numbers followed by 'n' refer to notes

Acadia National Park 85, *86*
Achilles tang (*Acanthurusachilles*) 119
adaptive capacity, in mountain tourism 40
adaptive management 13–15, **14**
adjusted standardized residuals (ASR) 88
alpine tourism *see* mountain parks, tourism in
animal welfare 120, 127, 128, 129, 132
Arthington, A. H. 3
ASR *see* adjusted standardized residuals (ASR)

Bangor International Airport (BIA) 85
Benson, C. 101, 109
BIA *see* Bangor International Airport (BIA)
Big Island (Hawai'i) 126
Boon, P. I. 3
Braun, V. 121
Bryce Canyon National Park 60, 63–4, 70, 72, 73–4
Burdisso, T. 67
Burton, I. 51n15

Canadian National Parks 34
Canyonlands National Park 64, 68, 70, 72, 73
Capitol Reef National Park 64, 70, 72–3, 75
Carrillo, C. M. 32
Ceron, J.-P. 18
CGE *see* computable general equilibrium models (CGE)
Chen, C.-W. 110
Cheung, C. 85
Chi, Y.-Y. 101
chi-squares 88
Chizinski, C. J. 32
Clarke, V. 121
Clay, E. 101, 109
climate change, perceptions of 90–1
climate extremes, in mountain tourism 46–8
climate trends, in mountain tourism 44–5
climate variables 5, 14, 59–60, 61, 65–6, **66**, 67–8, **69**, 70, **71**, 72–6

Climatic Research Unit (CRU) 65
Colorado Plateau 63–4
Colorado River Outfitters Association 51n3
community scale analysis 105
computable general equilibrium models (CGE) 101
costs, of tourism *see* tourism, costs and benefits in
Cramer's V 88
Creole Music Festival 104–5
Croom, A. M. 130
cross-sectional dependence (CSD) 67
CRU *see* Climatic Research Unit (CRU)
CSD *see* cross-sectional dependence (CSD)

Dawson, J. 18
Department of Land and Natural Resources–Division of Aquatic Resources (DLNR-DAR) 119, 130
de Urioste-Stone, S. 5, 81
disaster management *see* tourism disaster management
DLNR-DAR *see* Department of Land and Natural Resources–Division of Aquatic Resources (DLNR-DAR)
Dubois, G. 18
Duke, E. A. 1

EIS *see* Environmental Impact Statement (EIS)
environmental change 2, 6, 37; costs and benefits of 10–27; and disaster planning 111; effects on tourism 84
environmental engagement, tourism and 84–5
Environmental Impact Statement (EIS) 125
extractive resource use 127, 128, 132–3; non- 131

Falk, M. 61
Faulkner, B. 102, 105, 107, 110–11

Fellows, C. S. 3
Finland, tourism business in 5, 10–27
Fischer, F. 129
fish replenishment areas (FRAs) 118, 129
Fisichelli, N. A. 60
FRAs *see* fish replenishment areas (FRAs)
Freitas, C. 37, 51n2

Gabe, T. 5, 81
Gayle, R. 5, 58
GCC *see* global climate change (GCC)
GDP *see* Gross Domestic Product (GDP)
GEC *see* global environmental change (GEC)
Ghil, M. 101
Glacier National Park 46
global climate change (GCC) 11, 12, 13–14, 21, 25
global environmental change (GEC) 11, 12, 14, 21, 24
Grand Teton National Park 40, 42, 45, 46–7, 48, 51n12
Gross Domestic Product (GDP) 4, 26, 82, 101, 103, 111

Hadwen, K. J. 3
Haigh, T. 32
Hallegatte, S. 101
Hawai'i, aquarium trade in: animal welfare 120, 127, 128, 129, 132; culture 133; economy 130–1; emergent themes 123–4; ethics 121; extractive resource use 132–3; latent thematic analysis 121, **122**; legislation 130; limitations 134; overview 118–19; rank 128; rationale 119–20; regulations 129–30; results 122–8, **122**; social benefit and environment 131–2; subthemes 124–8; sustainability 129; thematic summary 133–4; user conflict 131
Hayes, M. J. 32
HB 873 hearings 120, 123, *123*, **124**, 125
Hewer, M. J. 83, 93
high-in-demand species, in aquarium trade 124
Hurricane David 101
Hurricane Hugo 101
Hurricane Katrina 101
hurricanes 99, 100, 101, 103, 105, 111
Hystad, P. W. 102

IE *see* Integral Ecology (IE)
infrastructural benefits 20
Integral Ecology (IE) 119
Integrated Resource Management Applications Portal 64
Intergovernmental Panel on Climate Change (IPCC) 5, 11, 37
IPCC *see* Intergovernmental Panel on Climate Change (IPCC)

Jedd, T. M. 5, 32
Jungle Bay Resort ('JBR') 105, *106*, **109**

Kajan, E. 5, 10, 100
KAPU system 128
Karl, M. 6, 98
Kates, R. W. 51n15
Keller, P. C. 102
Koontz, L. 51n13

Lamborn, C. C. 5, 58
Lapland (Finland) 15, 22, 23, 24, 25
latent thematic analysis 121
Law, R. 85
Lee, Y.-F. 101
Levine, A. 130
Likert scale 88
Lindell, M. K. 102
Linliu, S.-C. 101
lionfish (*Pterois volitans*) 118
Lise, W. 67
Liu, T.-M. 60
local climate data, in Utah 65–6, **66**
Loomis, J. B. 38, 52n17, 60

management: adaptive 13–15, **14;** sufficient 125, 129; *see also* tourism disaster management
Maine, tourism in 5, 82, 84, 85–94, **88**, **89**, **90**, **91**, **92,**
Mair, J. 107
'Marine Protected Areas' (MPAs) 125
marine tourism 118
Mather, S. 73
McBoyle, G. 18
McCool, B. N. 111
McKercher, B. 85
Michael, F. V. 101
micro-level analysis model (MLAM) 6, 107–9, *108*
Mieczkowski, Z. 83
Mighty 5 national parks 62–3, *63; see also individual parks*
MLAM *see* micro-level analysis model (MLAM)
Morgan, R. 83
mountain parks, tourism in: adaptive capacity in 40; climate extremes in 46–8, *46*, **47**; climate trends in 44–5; data selection and processing 42–4, *43–4*, **43–4**; economic significance 38–9, **39**; non-climatic sensitivity 39–40; overview 33–4; physical changes 37–8; study sites 40–2, *41*; visitation outcomes and economic impacts 45–6, *45*, **46**; vulnerability study 36–40; weather and climatic effects on 34–6
MPAs *see* 'Marine Protected Areas' (MPAs)

National Park Service (NPS) 33, 38, 42, 45, 60, 62, 64, 66, ; *see also specific parks*
nature-based generalists 89–93
nature-based tourism: climate change perceptions 90–1, **92**; data analysis 87–8, **87**,

88; and environmental engagement 84–5; in Maine 84; overview 82; segmentation 88–9, **89**; study site 85–6, *86*; survey design and sampling procedure 86–7; visitor profile 89, **90**; weather and climate 82–4; weather sensitivity 89–90, **91**
NIMBY *see* Not in My BackYard (NIMBY)
non-climatic sensitivity 39–40
non-extractive resource use 131
non-response bias 86, 88
Not in My BackYard (NIMBY) 121
NPS *see* National Park Service (NPS)

Ooi, N. 1
O'Leary, J. 1

panel-specific estimation 70–2, **71**
Pelling, M. 102, 107, 111
Pesaran, M. H. 67
PET *see* potential evapotranspiration (PET)
Porter, B. A. 117
Post Hoc 88
potential evapotranspiration (PET) 38
Prideaux, B. 85

Rasmussen, T. N. 101
recreation park visits, monthly 64–5, *65*
reformed aquarium collectors 125, 130
Richardson, R. B. 38, 52n17, 60
Ritchie, B. W. 105, 107, 110–11
'The Road to Mighty' ad campaign 74
Robertson, A. W. 101
Rocky Mountain National Park 84
Rossiter, J. S. 130

Saarinen, J. 5, 10, 100
Saariselkä, business in 15–16, 18–23
Saffir Simpson scale 103
SAIA *see* Sustainable Aquarium Industry Association (SAIA)
Sangiácomo, M. 67
Schmude, J. 6, 98
Schwaiger, K. M. 6, 98
Scolobig, A. 111
Scott, D. 18, 60
SIDS *see* small island developing states (SIDS)
Skerrit, Roosevelt 103
Skidmore, M. 107
small island developing states (SIDS) 99, 103
Smith, J. W. 5, 58
social benefit, and environment 131–2
SPEI *see* Standardized Precipitation and Evapotranspiration Index (SPEI)
Standardized Precipitation and Evapotranspiration Index (SPEI) 37
Stern, N. H. 26
Stevens, T. H. 33
Stevenson, T. C. 131
sufficient management 125, 129

SurveyGizmo software 87
sustainability 11, 25, 26, 27, 124; in disaster recovery planning 102; Hawaiian aquarium trade 119, 120, 123, 124–5, 126, 128, 129–30, 131, 133, 135
Sustainable Aquarium Industry Association (SAIA) 119
sustainable communities 102
sustainable development 13, 103
sustainable tourism 103, 105
Swigart, J. 32

Tailored Survey Design Method 87
TAMS Analyzer 121
Taylor, B. 3
TCI *see* tourism climate index (TCI)
Tervo-Kankare, K. 5, 10, 100
thematic summary 133–4
Thomas, C. C. 51n13
Thomas, D. S. K. 34, 37
Thornwaite, C. 38
Tissot, B. N. 119, 131
Todd, G. 73
Tol, R. S. J. 67
tourism climate index (TCI) 83
tourism, costs and benefits in: adaptation perspectives 13–15, **14**; changes in natural environment 18; changes in socio-economic environment 18–20; environmental change benefits 20–1, **20**; environmental change costs 21–3, **21**; in Kilpisjärvi and Saariselkä **17**, 18–23, **19**; overview 11–13; study design, methods and data 15–18, *15*, **17**
tourism disaster management: literature review 100–2; micro-level analysis model (MLAM) 107–9, **109**; overview 99; research model and data 105–7; study area 102–3; tropical storm Erika 103–5, **104**, *106*
tourism, in Maine *see* Maine, tourism in
Toya, H. 107
Trail Ridge Road 34
Tropical Storm Erika 6, 103–5, 109, 111
tropical storms 6, 99, 100, 103–5, 109, 111
Tsai, C.-H. 101, 110

UNCCC *see* United Nations Framework Convention on Climate Change (UNCCC)
United Nations Framework Convention on Climate Change (UNCCC) 11
United States Census Bureau 122
US dollar index 66
US National Park Service units 60
Utah Office of Tourism 74
Utah, climate and visitation in: analysis 67; characteristics 63–4; climate variables 72–5; common correlated effects estimation 70, **71**; description 60–1; descriptive statistics 68–70, *68*, **69**; local climate data 65–6, **66**;

Mighty 5 national parks 62–3, *63*; monthly recreation visits 64–5, *65*, **65**; overview 59–60; panel-specific estimation 70–2; US dollar index 66; weather sensitivities 61–2

Vail Resorts 51n3
Vikulov, S. 110
Viner, D. 73
Visitor Use Spending report (NPS) 42

Wall, G. 101
Walsh, W. J. 131
Walters, G. 107
Waterton Lakes National Park 60

weather: and climate 34–6, 82–4; sensitivity 61–2, 89–90
Weiskittel, A. 5, 81
Welsch, E. 121
West Hawai'i Fisheries Council 119
White, G .F. 51n15
Wilkins, E. 5, 58, 81
Wolsink, M. 121
Wu, T.-C. 101

Yellowstone National Park 40, 42
Yu, G. 61

Zavareh, S. 6, 98
Zion National Park 64, 68